Pelican Books

The Space Busine[illegible]

Peter Marsh was born in Ju[illegible] [illegible]ty Grammar School and Notti[illegible] [illegible]ere he gained an honours degree in chem[illegible] [illegible] went on to become a journalist, and has worked with the *Evening Post* (Luton), *Building Design* (an architectural publication), *New Scientist*, and the *Financial Times*, where he has been a technology reporter since December 1983. His previous books are *The Silicon Chip Book* (1981), *The Robot Age* (1982) and *Robots* (1983). He has written about the space industry for *New Scientist* and the *Financial Times* since 1980.

Peter Marsh lives in Balham, South London, with his wife and son. His interests are playing football and gardening.

For Christopher

Contents

List of Figures

☆

List of Tables

☆

Acknowledgements

The volumes and reports listed in Further Reading acted as source material for much of the book. Other material came from information gleaned personally from people in the world's space industry. Besides these people, I owe thanks to three other groups: to the staff of the British Library, where I was able to read many of the books listed; to the staff of *New Scientist* for their help and support during the time I worked with them; and, finally, to Nikki, my wife, and to Rodney Buckland for reading parts of the manuscript and for making helpful suggestions on how to improve it. Thanks to Ray Ward for checking the figures in Table 13.

1 ☆ The Exploitation of Space

Of all the changes that have affected the world over the past thirty years, one is particularly significant. Space – that volume of the universe that begins when the Earth's atmosphere starts to peter out about 160 km above our heads – has ceased to be something purely to wonder at. Instead, it has become an entity to be explored and, still more so over the past few years, one whose characteristics are to be exploited.

Space today forms a medium in which to place satellites, for telephone communications, for observing the land and oceans and for monitoring the weather. Other satellites provide navigation services for ships and a relay point from which to channel TV into our homes. Space is extremely important for military forces, and is becoming still more so. It already houses a range of craft that form a crucial part of the military infrastructure of the superpowers; for instance satellites are used to send data to weapons systems or between command posts, and other vehicles are continually spying on the military hardware of their enemies, probing out strengths and weaknesses. The vast expanses of vacuum beyond our atmosphere could also soon act as a home for a new range of space weapons built by the USA and the USSR either to 'kill' opposing satellites or, still more ambitiously, to neutralize missiles with beams of high-energy atomic particles or laser light.

A less threatening side of space activities concerns the use of new generations of orbiting bases in which people reside permanently. The Soviet Union has made the greatest progress in this direction and has launched a series of rudimentary space stations, called Salyuts, in which cosmonauts have lived and worked for up to eight months at a time. The USA, following an announcement by President Reagan in January 1984, is to build its own space station by the 1990s. The orbiting structure, to consist of laboratories and work spaces for factory operations, is intended also to be for use by other countries, predominantly Western Europe and Japan.

The incentive which drives both the USA and the USSR to develop space stations is the likely high profitability of working in space. Inside a craft in revolution about the Earth, gravity is reduced to a tiny fraction of that experienced on our planet. In this environment, industrial operations, for example producing new kinds of materials that are difficult to make under Earth's gravity, may become highly lucrative. Several companies, mainly in the USA, are interested in researching the possibilities of operating fully-fledged space factories in the 1990s or early part of next century. Among the products could be new kinds of drugs and alloys and very pure crystals of materials such as silicon.

Space looks likely to be a habitat admirably suited not so much for human beings as for robots. By the end of 1984, 155 individuals from fifteen different countries had ventured out of the atmosphere. Between them, they had made 102 separate manned missions into the cosmos, including, of course, the six pioneering visits to the Moon by US astronauts. However, space is no place for people. It is too unpleasant and threatening and, above all, too expensive.

The work of men and women in space in the decades to come will be left largely to mechanical artefacts containing a high degree of advanced electronics, so giving the machines a level of autonomy and power to make decisions of their own. Development of these sophisticated robots has already started in some of the world's leading industrialized countries. The work will stretch the capabilities of engineers and scientists engaged in applications of 'intelligent' computers which have a reasoning ability approaching that of the human brain.

Space, it almost goes without saying, is an international arena. Virtually every country is dependent on what happens in space, whether because its economy relies on transfer of information via satellites or because it gleans vital data, perhaps about its own croplands or mineral deposits, from photographs of the Earth taken at an altitude of a few hundred kilometres by remote-sensing satellites.

The trend towards a greater international involvement in space activities is strengthened as more countries develop their own expertise in space technologies. The USA and USSR, the leaders in the exploration of space in the 1950s and 1960s, are still the major space powers. In a 'second league' are countries such as Japan, Canada and Western Europe. The Western European countries clubbed together in the mid-1970s to form the eleven-nation European Space Agency (ESA), a co-ordinating body for space activities. The members are France, West Germany, Italy, Britain, Spain, Sweden, Switzerland, Denmark, the Netherlands, Belgium and Eire. A 'third league' comprises the countries that either operate, or are planning to operate, satellites, or which are developing their own launchers. Such nations include Pakistan, Indonesia, India, Brazil, China, South Korea, Australia, Saudi Arabia and Mexico. Finally there are those countries which rely on satellite data of one kind or another. Some of the nations in this category have formed regional groupings to make the most of their capabilities in space technology. For instance, several African countries co-operate on remote-sensing technology, and Latin American nations, led by Chile, Argentina and Colombia, have talked about forming their own South American space agency.

An area in which there is significant international co-operation concerns scientific studies. From the early days of space exploration, a principal motivating factor in developing space technologies has been to discover new knowledge about either the Earth or the cosmos. Space affords a good vantage point from which a satellite can be employed to take measurements related either to our planet or its atmosphere or to bodies such as the stars and other planets of the solar system. The countries of both the industrialized and developing worlds have co-operated on many occasions in operating space vehicles, or in gathering data from them via ground receiving stations.

Ventures in space stations will also aid scientific understanding, not just of materials and industrial processes but of disciplines such as life sciences and astronomy. In space flight, with the force of gravity stripped away, scientists may obtain new understanding of phenomena which, on Earth, are masked or distorted by gravitational pull. Activities which will benefit from this kind of work could include studies of the growth mechanisms of plants and human physiology. Research of this sort will often be the result of shared ventures between different countries. A good example was the first flight of Spacelab at the end of 1983. Spacelab, the world's first re-usable space workshop for scientific experiments, was built by the nations of ESA and was carried into orbit by a space shuttle built by the USA.

Competition between governments and between business organizations will be a feature of the next few decades in space. A principal area of competition is in the offering of services to put satellites into orbit. Historically, the USSR and the USA have been dominant in this area with their development of expendable rockets to do the job. Using this mechanism, the Soviet Union put into space the first artificial satellite (in 1957) and the first person (in 1961). Other countries, however, have developed a similar capability. France, Britain, Japan, China and India have all launched satellites with their own rockets. Commercially, the main rival to US vehicles in offering a service to launch satellites is Ariane, a rocket developed by the European Space Agency and now sold by Arianespace, a semi-private company in which French interests hold the dominant shareholding.

The US space shuttle is the most advanced of the launch vehicles offered by US concerns. It is the world's first re-usable space vehicle – having carried into orbit a payload such as a satellite, the craft returns to Earth like an aeroplane. A total of four shuttle vehicles (called orbiters) should be operating by the end of 1985 under the responsibility of a government body, the National Aeronautics and Space Administration. Further in the future, however, the shuttle fleet may be operated by a private body. Such a fate has already befallen some of the expendable rockets that the US government developed in the 1950s and 1960s. For example, Transpace Carriers is attempting to sell launches on the Delta rocket, and General Dynamics is marketing the Atlas Centaur. The move towards 'privatizing' space technology, developed initially by the government, is a strong feature in the USA under the administration of President Reagan. For instance, the US government is selling the Landsat remote-sensing satellites, which provide data about land and sea areas, to the private sector.

Competition between private concerns will be especially strong in the area of communications satellites. The commodity in contention here is less tangible than physical items such as metals or energy. The commodity is information. Many of the cultural and commercial battles of the future will be fought over it. A satellite in orbit round the Earth acts as a convenient relay stage for transferring information in the form of radio waves from one place on the planet's surface to another. The information can be in the form of voices (as in telephone calls), banks of computer or electronic data, or images in TV programmes. Satellites act not only as passive relay vehicles but as active harvesters of information. With sensors such as TV cameras or infrared scanners, these remote-sensing satellites can glean information about mineral deposits or trace meteorological patterns.

Three-quarters of the satellites that mankind has sent into orbit have had military purposes. Indeed, without military projects the space age would have started later, and its development would have been much slower. Rockets were devised initially to carry weapons. Armed with a bomb and cruising through space in a ballistic trajectory, the rocket seemed to the USA and the USSR during the 1950s to encompass a useful piece of hardware for striking at an enemy, and a position in space gives new meaning to the military phrase 'the high ground' – a vantage point from which one adversary can gain an advantage over another. The permanent bases that the USA and the USSR will establish in space will undoubtedly have military uses. Hermann Oberth, a pioneering German rocket engineer, recognized this when he wrote in 1957: 'Whenever work on a space station is started it will undoubtedly be for military reasons. The crew of the first space station will certainly wear uniforms, most likely a special fatigue suit.'*

The military aspects to space technology have taken on a new dimension with President Reagan's announcement in 1983 of a 'Star Wars' programme to develop space-based beam weapons to shoot down incoming Soviet missiles in flight. The project, more properly called the Strategic Defence Initiative and costing $26,000 million in federal research cash to 1989, could lead to a working antimissile system by around 2000 that covers the whole of the USA. The US government says this system would be essentially defensive – it would not be used to mount offensive strikes against military or civilian targets in the Soviet Union. But the existence of the research project could impel the USSR to increase its own efforts to find ways of beating a Star Wars antimissile barrier, thus creating a new impetus to the arms race. The Star Wars project is concurrent with

* Details of books mentioned in the text are given in Further Reading.

efforts by both the USA and Soviet Union to build antisatellite weapons to shoot down the opposing country's space vehicles, communications craft used for military purposes for example.

The signs are that President Reagan's enthusiasm for the new space weapons is not shared by some of the USA's partners in the NATO Alliance. Britain, France and West Germany all hinted in the closing months of 1984 that they had reservations about the development of such hardware on the grounds of its destabilizing effect. It appears that the issue could cause splits in NATO in the next few years over the whole area of armaments policy.

The aspects mentioned so far will all be discussed in some detail in later chapters. This, I hope, will enable the reader to gain an insight into the extent to which the applications of space technology are affecting, or will affect, the lives of ordinary people.

First, a few basic facts about space itself. The Earth is surrounded by an atmosphere of gas, mainly oxygen and nitrogen but containing numerous trace molecules such as methane and oxides of carbon and nitrogen. The atmosphere gradually becomes thinner with increasing distance from the planet; in practical terms, its limit is normally taken to be at an altitude of about 160 km. 'Space' is everything beyond this limit. Much of space is, therefore, by definition either vacuum or near vacuum. However, according to this definition, space also contains a great many solid bodies – the other planets of the solar system for instance.

This book is mainly concerned with the applications of 'near space' – that part of space within relatively easy reach of our planet. None the less it is worth examining for just a few moments the immenseness of space as a whole. The body closest to the Earth is the Moon, at a distance of 380,000 km. The Earth travels round the Sun, which is 150 million km distant. Of the other eight planets in our solar system, Mercury and Venus are nearer the centre of the solar system, 58 million and 107 million km respectively from the Sun. The other planets are Mars, Jupiter, Saturn, Uranus, Neptune and Pluto. They are hundreds, or even thousands, of millions of kilometres from the Sun.

A useful exercise is to reduce these distances to a more familiar scale. If the Earth is represented as a ball a couple of centimetres in diameter, the Sun is a globe 4 metres across and nearly 250 metres away. The Moon is a small pea 1 metre from the Earth. Mars is 340 metres beyond the Sun, Jupiter 30 cm in diameter and a little over a kilometre away, with Saturn 2½ km, Uranus 5 km, Neptune 7 km and Pluto 10 km distant. Then, nothingness for some 70,000 km, at which point the nearest star would appear.

The galaxy in which we live contains some 10^{11} star networks similar to our solar system of the Sun and planets. (Written out in full, this number is 100,000,000,000.) That part of the universe which astronomers can detect with telescopes contains another 10^{9} galaxies similar to our own. So that makes a total of 10^{20} stars just within that part of creation which we can visualize. The figure is impossible to comprehend. Arthur C. Clarke has pointed out that a telephone directory with all these numbers would require more pages than all the books ever printed.

Looking at this in another way, imagine that a super-fast computer set out to digest this number of instructions. Such computers have no trouble assimilating 10 million instructions per second. Even at this fearsomely high speed, the computer would take 300,000 years to complete its task. If it had started at the time Stonehenge was built, the machine would be only one-hundredth of the way through its task.

Superimposed on a model of even that part of the universe within striking distance of the Earth, the part in which man can comfortably live is tiny indeed. Without protection or life-support systems, men and women can venture only minute distances from the Earth's surface – either up or down. A few kilometres towards the centre of the Earth and the air becomes too hot and the pressure too high to support life, human or animal. The deepest mines go no more than 4 km below the surface, a distance that, on a horizontal scale, a child of seven would find laughably simple to cover on an afternoon's outing. Moving upward from sea level, at about 3 km oxygen pressure falls off to the extent that most people find breathing difficult. At this point, for people unaccustomed to high altitudes, even the mildest exercise is troublesome. At about 15 km, not enough oxygen is available to support life. A little further from the Earth, blood starts to boil as a result of the fall in pressure.

Our atmosphere not only provides the air which we breathe, it also shields us against cosmic rays that would harm the skin. At a height of about 40 km, the protection is stripped away. At 90 km the sky becomes black. This is because the number of gas molecules from which light can be scattered has fallen so low as to be negligible. Even here, though, space is certainly not empty. At altitudes greater than 120 km, a space traveller would encounter meteorites, objects of a very solid nature. These projectiles rarely get any nearer the Earth's surface; below this point, gases in the atmosphere are dense enough to burn them up. A little further, at around 160 km, and the borders of space begin. Air molecules become so scarce that solid bodies encounter no measurable resistance to movement; so support for aeroplanes in the form of aerodynamic forces no longer exists. (The practical limit for aircraft is 50 km.)

Molecules of gas do not disappear altogether. Even in the depths of space, a few particles have managed to wrest themselves away from the mantles of gas surrounding a planet such as the Earth. Each cubic centimetre of air at sea level contains about 10^{18} gas molecules. The comparable figure for 'space' is between 1 and 100.

We have to thank gravity for preventing the Earth's valuable atmosphere from straying away. Gravity is the force of attraction between every two bodies in the universe. It is directly proportional to the product of the bodies' masses and inversely proportional to the square of the distance between them. Only one form of gravity is important to most people – the attraction between the Earth and the objects on or near it. The mass of the Earth is so large that the gravitational forces due to this factor swamp all the other forms of gravitation, which is why, even though the armchair and the table in your sitting room attract each other, this force is insignificant compared with the gravitational tug of the Earth that keeps both items firmly anchored to the planet's surface. The same force of gravity hugs to the Earth the molecules of the atmosphere. This force is at the heart of the problems encountered when attempts are made to launch objects such as satellites or rockets away from the planet.

The gravitational attraction of the Earth never fades away – it extends to infinity; but it has very little effect a couple of million kilometres out into space. Long before an object in a journey from Earth reached this point, it would come under the influence of the other massive bodies in the solar system. Of these the Sun has the biggest effect; the Sun's gravitational pull keeps the Earth, and the other planets of the solar system, firmly in place in orbit around it.

Another set of influences is almost as important – those of governments. The governments of the industrialized world have seen space activities as important from several different viewpoints – military, political and economic. The two big space powers, the USA and the Soviet Union, every year spend between them on space activities (in roughly equal proportions) about $30,000 million – a sum roughly equal to the gross domestic product of a country of the size of Portugal. About half of this spending is on military work.

Prestige is important in space. The USA and the USSR started out in space technology (after the defence considerations) because of the political kudos gained by feats of daring outside the Earth's atmosphere. The Soviet Union reaped much political mileage from its launch in 1957 of Sputnik 1, the first artificial satellite. The newspaper *Soviet News* commented at the time:

> The successful launching of the first man-made satellite makes a tremendous contribution to the treasury of world science and culture . . . The artificial Earth satellites will pave the way for interplanetary travel and it seems that the present generation will witness how the freed, conscious labour of the people of the new socialist society turns even the most daring of mankind's dreams into reality.

In 1961, President Kennedy said the USA would put a man on the Moon before 1970. The statement foreshadowed what happened eight years later. In a book published in 1961, Wernher von Braun, one of America's leading rocket engineers, summed up why the USA had to make a lunar landing before the Soviet Union. He wrote in *First Men to the Moon*: 'A country from which a major proportion of mankind expects dynamic leadership has no choice but of either taking up the historic challenge of the day or of stepping down from the position to which fate has lifted it.' The efforts to reach the Moon were as much an act of political will-power as a technological achievement.

Many people criticize the large sums spent by governments on space projects. They do so on the grounds that the cash could be better spent on work to better the lives of people on Earth in more direct ways – or that the projects themselves are instigated not so much to provide economic returns but to bolster the reputations of people working in state space agencies.

The plan by the US National Aeronautics and Space Administration to spend $8,000 million by 1992 on a manned space station has attracted this kind of criticism. An important report on the subject in late 1984 by the US Congress Office of Technology Assessment said the project was far too ambitious and costly. Equivalent goals in establishing a manned 'infrastructure' in space could be accomplished far more cheaply, said the report's authors, for instance by extended flights of the space shuttle or by building less expensive unmanned platforms.

According to other critics, NASA had used its hefty lobbying power to persuade President Reagan of the merits of the station in order to give its large corps of engineers a meaty project to work on during the coming decade.

The OTA report was adamant that the space station 'cannot now be fully justified on scientific, economic or military grounds or combinations thereof'.

It would be wrong, however, to think of developments in space technology as the result simply of initiatives by self-seeking governments or by greedy commercial organizations. There is also a spiritual side to space exploration. For centuries, people have inquired on a theoretical or religious level into the nature of the heavens; the immense tracts of the cosmos are regarded with awe. The possibilities of travelling into space and finding other forms of life continue to fascinate millions – as is shown by the reaction to popular films on the subject.

Space activities hold out some promise, in theory at least, of bringing together people on Earth. In 1956 Pope Pius XII spoke of this:

> While up to now Man has felt himself banned on Earth, so to speak, and has had to be satisfied with fragments of information which come from the Universe, it appears now that the opportunity has come to put aside the barriers . . . Man's wish, peacefully to conquer the Universe, should contribute to deepening the feeling of community and solidarity, so that all should have the feeling of belonging to the great family of God, children of the same Father.

The historical element to space activities is important, and this is the subject of the next chapter. Although many aspects of space exploitation today are novel – the operation of the space shuttle is just one of them – in many ways the development of space technologies follows a logical progression that first became apparent some decades ago. A report on civilian space policy by the US Congress's Office of Technology Assessment observed in 1982:

> Perhaps one of the most remarkable aspects of the way the space program has developed is the fact that the opportunities and areas of activity in the space program have not changed appreciably over a quarter century. The developments in space applications, the mission opportunities in science and the manned space exploration program have largely followed the scenario laid by the early advisers and space proponents – if anything they have failed to equal the imagination and vision of these early projections.

The space technologists of today owe a large debt to the people who showed the way into space – the theorists, the engineers, the fanatics, the showmen . . . and the storytellers.

2 ☆ The Propulsion Pioneers

Beads of sweat stood out on Ivanov's red face. His hair was in a tangle; his eyes were shining with bliss and weariness; he seemed like a man unnaturally inspired. 'Four days and we can be on the Moon; outside Earth's atmosphere in a few minutes; interplanetary space a hundred days from now!' he suddenly blurted out. 'You're delirious,' said Newton, the Englishman.

This little cameo marks the opening of a space fantasy, *Beyond the Planet Earth*, first published in 1920. The author, a Russian schoolmaster named Konstantin Tsiolkovsky, was more than just a relater of stories. His tale of an international venture into space was based on rigorous scientific theories that he had himself put forward. Later the theories formed the basis of much of mankind's effort at making space travel practicable.

In his book, Tsiolkovsky tells how six scientists, led by the Russian, Ivanov, but including representatives from Germany, France, the USA and Italy as well as the initially sceptical Englishman, get together to devise a rocket and take it off into outer space. The text has a delightfully idealistic tone which may seem out of place in the cynical 1980s. Notice, for instance, this ringing speech by Newton, the Englishman, on the return of the space pioneers from their first voyage. Touching on the subject of space colonies, Newton postulates:

> When the colonies are sufficiently numerous, they will develop their own industry out there; they will start to build their own homes and will cease relying on support from the Earth . . . In return, many of our descendants will find shelter, happiness and complete moral contentment out in the skies.

Newton continues:

> What human genius can predict how the settlements beyond Earth's orbit will be, a thousand or a million years from now? . . . Can we foresee

what they will achieve, how their industry and science is going to develop, how man himself will be transformed out there?

These questions are only marginally less difficult to answer now than they were in 1920. The fact that Tsiolkovsky was asking them years before mankind had any practical experience of space travel marks him out as one of the true visionaries in this subject.

In *Dreams about the Earth and the Sky* (1895) and *Exploration of the Universe with Rocket-Propelled Vehicles* (1903), Tsiolkovsky laid the ground for what was actually to happen several decades later. The Russian theorist suggested using liquid propellants in rockets, of the kind that engineers in the 1960s harnessed in sending vehicles to the Moon. He showed how rockets could be controlled by tilting their exhaust nozzles and by metering the flow of the propellants to the rocket engine. Tsiolkovsky also recommended building spaceships made with double walls for protection against excessive heat or cold and to shield the passengers from meteorites. In one of his books, the schoolmaster even discussed how artificial satellites would operate.

If anything, Tsiolkovsky was a little cautious about predicting when the events he described would take place. He said the first 'space walk', accomplished by one of his own countrymen in 1965, would not take place until the twenty-first century. He was also careful to set the events in *Beyond the Planet Earth* no earlier than the year 2017.

Tsiolkovsky was one of a long line of Soviet engineers and theorists who contributed much to rocketry. There is a record of a rocket factory in Moscow as far back as 1680. The projectiles were used as weapons by the Russian army and navy. In 1817, an inventor in St Petersburg called Alexander Zasyadko demonstrated such a rocket with a range of 3 km. Nine years later, a permanent rocket establishment opened its doors in the city; its job was building weapons for the army. Later that century a scientist called Nikolai Kibalchich put forward a detailed proposal of how a piloted rocket-propelled vehicle would operate.

Nineteenth-century Russians, however, were not the only ones displaying a fascination for rocketry. Indeed this is something that has kept people all over the world occupied for centuries. G. Edward Pendray, a rocket enthusiast himself, summed it up well in 1947 in his book *The Coming Age of Rocket Power*:

There is something about rocket power which transcends the bleakly mechanical aspects of the subject and changes its followers into missionaries . . . Perhaps . . . it is the lurking chance . . . that by rocket power we shall actually someday be able to visit and explore worlds other than our own. Whatever it may be, those of us who have spent years in the study and development of rockets have acquired an emotion about them which is almost religious.

According to most history books it was the Chinese who first hit on the idea that an object could thrust itself forward by producing an explosive force in the opposite direction – the principle behind the rocket. To provide the explosion the Chinese used a mixture of charcoal, sulphur and saltpetre (it became known as gunpowder only several centuries later, when firearms were properly established). The powder was packed into a cylinder attached to an arrow. When ignited, the rocket flew off in the direction in which it was pointing. This was the application for the devices said to have been favoured by a certain Kiang-chin, governor of Kai-fung-fu, a town north of the Yellow River. In the year 1232, he ordered his soldiers to fire these weapons at an army of Mongolian invaders. The tactic certainly created terror – writers said the weapons were like 'thunder that shakes the heavens' – although the exact military effect of the rockets is unclear.

Other countries were to follow the Chinese lead in using the rocket for warfare. Historians have documented the use of rockets by Italian soldiers during the battle for the Isle of Chiozza in 1379. The French military tried out the devices during the defence of Orleans in 1429 and the siege of Port Andemer twenty years later. The British were on the receiving end of rockets when doing battle in India in the late eighteenth century. During the fight for the town of Seringapatam, the troops of the Tippoo Salaun of Mysore applied the devices with considerable success.

Stung by this, Britain produced its own rocket expert in the shape of Sir William Congreve. By 1806, he had made, at Woolwich Arsenal near London, some 13,000 rockets each with a range of 2 km. The British military first used the devices at sea, during the sieges of Boulogne and Copenhagen in 1806 and 1807.

Rockets had their peaceful side too. In the early 1800s, Claude Ruggieri, who came from a family of firework makers, tethered small animals to the devices and fired them across open fields near

Paris. He wanted to extend his experiments to human cargoes, but the French police, not wanting a mass of corpses on their hands, managed to restrain him.

At about the same time, a Cornishman named Henry Trengrouse was shooting rockets into the sea to find out if they could carry a line to boats. The devices could, he thought, have great application in rescuing people off Britain's stormy coastline. A fully-fledged rocket station for tending stricken vessels opened up on the Isle of Wight in 1826.

Another Briton, the singularly named Charles Golightly, received a patent in 1841 for a flying machine powered by a rocket motor. But then in 1875, spoil-sport legislators cut off the flow of British rocket developments in their prime. Parliament passed an Explosives Act, a law rigorously enforced, that forbade rocket testing and so Britain's band of technological enthusiasts had to turn to other pastimes.

In the United States in the last couple of decades of the nineteenth century, rocket-driven balloons were all the craze. In 1893, a US inventor named Sumpter B. Battey proposed what he called a 'machine-gun' mechanism for driving an airship. It is not recorded whether this project succeeded or whether it met the fate of another, much earlier venture in rocket transport conceived by a Chinese government official named Wan Hoo. Around 1500, he had the fantastic idea of building a rocket aeroplane comprising two large kites fixed to forty-seven powder rockets. But Wan Hoo was several centuries ahead of his time. The technology was simply too dangerous and he perished during the craft's first and only mission.

All these enthusiasts were using (though most of them would never have realized it) a principle first enunciated in 1687 by the English physicist Sir Isaac Newton and known as Newton's Third Law: to every action there is an equal and opposite reaction. The force imparted by a jet of gas or other material flowing from a rocket's exhaust is counterbalanced by a thrust pushing in the opposite direction. It is this thrust that propels the rocket. In a powder rocket, the only type available until the twentieth century, the jet comprises the gaseous products of the explosion created when gunpowder is ignited. A simple illustration of Newton's Third Law is what happens to a balloon that is filled with air and released. The air flows out of the hole in the balloon, which moves forward as a result. The jet engines of modern aircraft use a similar principle. They differ from rockets in that they require a continuous stream of air, which they take from the atmosphere. This serves as an oxidizing agent which, when mixed with a fuel such as kerosene or petrol and ignited, produces an explosive thrust. Rockets, by contrast, carry with them both a fuel and an oxidizing agent; in the case of gunpowder, saltpetre is the oxidizing agent and charcoal and sulphur the fuel.

Around the turn of the nineteenth century, rocketry left the realm of the theorist or the enthusiastic amateur. It became a subject for practical people. The appeal of rocketry was twofold. First, any vehicle that was to journey into space (where by definition there is no air that can be used as an oxidizer) needed a propulsion mechanism that carried its own oxidizing agent. Secondly, space pioneers were impressed by the very high propulsive forces that rockets can provide. These forces, so the mathematicians among the early rocketeers worked out, would be needed if ever vehicles were to escape from the terrific gravitational attraction of the Earth. These people quickly calculated that shooting objects up into the air with cannon, a technique advocated by nineteenth-century novelists, would not provide enough thrust for the job.

It was not long after this that the early experimenters in rocketry turned their attention away from gunpowder as a fuel. As anyone who has ever played with fireworks knows, gunpowder is tricky to control. Once ignited it will burn until it runs out. Short of quenching the explosion with water or some such agent, there is no way of moderating the combustion. A successful space vehicle, however, requires a fuel that engineers on the ground (or automatic equipment aboard the craft) can control. This is so that the spacecraft's thrust is varied to ensure that the vehicle leaves the ground in the first place and then travels in the way intended. Constructing a vehicle for space travel that uses only solid fuel is about as sensible as going for a ride in a car in which the accelerator pedal is stuck firmly to the floor.

The early engineers, therefore, looked to liquid fuels that could be fed into the combustion chamber of a rocket in carefully metered quantities controlled by valves. Once in the chamber, ignition could take place to provide a thrust of known power and duration. After this, if it was so desired, the flow of fuel could be switched off. Examples of liquid fuels are methane, kerosene, petrol and alcohol. In each case they need an oxidizing agent such as liquid oxygen. More modern fuel/oxidizing agent combinations are liquid hydrogen/liquid oxygen and nitrogen tetroxide/unsymmetrical dimethyl hydrazine (UDMH).

Three groups of inventors – in the Soviet Union, the USA and Germany – experimented with early liquid-fuel rockets. These people played important roles in devising the first practical space vehicles. Russia in the 1920s was the centre of enormous interest in the practical aspects of rocketry. No doubt this had much to do with the work of early theorists such as Tsiolkovsky and the country's long and honourable traditions in this area. The enthusiasm transcended scientific or professional disciplines. In 1928, the Leningrad Institute of Railway Transport Engineers formed a subsection for its members interested in space travel. At around the same time, the Soviet Union's Association of Inventors held what was billed as the world's first exhibition of spacecraft design.

More consequential was the formation in 1928 of the Gas Dynamics Laboratory near Leningrad. The following year it turned its attention from solid-propellant rockets (though this work still went on, mainly for use with short-range weapons for the battlefield) to vehicles that used liquid fuels.

At first the laboratory built these devices for the air force to provide extra thrust for propeller-driven aeroplanes. Then in 1932, the Gas Dynamics Laboratory provided the liquid-propellant engines for a rocket that took off vertically to a height of 80 metres.

The vehicle itself was designed by another team of engineers at an army rocket establishment in Moscow. The head of the group was a certain Sergei Korolev, a name that within three decades was to create quite a stir. He designed the rocket that put Sputnik 1, the first satellite, into orbit in 1957 – of which more later.

However, long before this event, exciting developments were taking place several thousand kilometres away, on a farm in rural Massachusetts. An American inventor first demonstrated to the world that a rocket powered by liquid fuel could lift off the ground. Practical demonstrations of this technology understandably stretched the resources of the engineers of this time. They had to build sophisticated pipework to carry the fuel and oxidizing agent to a chamber where ignition takes place. They had to time very carefully the moment when ignition starts. They had to control the passage of the liquids with sets of valves. On top of this was the hazardous nature of the propellants which could leak out of their containers at any time, possibly causing an explosion.

The American pioneer was Robert Goddard, a stubborn, shy and sickly man obsessed with building machines which would, as he correctly reasoned, lay the foundations for taking people into space. The date of Goddard's greatest triumph was 17 March 1926. Experimenting at his aunt's farm in Auburn, Massachusetts, Goddard fired his rocket to a height of 12 metres. The device, powered by petrol and liquid oxygen, travelled 54 metres in 2.4 seconds. Much of Goddard's later career was dogged by ill luck and by his penchant for annoying people. In 1929, another rocket launch from the farm caused an uproar among neighbours. They reasoned that anyone occupying himself obsessedly with such activities was off his head. The police were called, Goddard was duly admonished and, thereafter, he had to resume his experiments at a nearby military base where it was assumed the authorities would keep an eye on him.

Much of the sentimentality that surrounds Goddard's life dates from an event that occurred in 1899, when he was seventeen. According to Goddard's recollections later, he climbed a cherry tree and had a vision of ascending to Mars. He wrote: 'I was a different boy when I descended the tree from when I ascended. Existence at last seemed very positive.' Reverberations from this curious episode cropped up during Goddard's life. From 1899 on, he made 19 October a special 'anniversary day' in recognition of his boyhood experience. The inventor had other visions. In his diary for 8 August 1915 he recorded a dream of a trip to the Moon.

In the 1930s Goddard found a sponsor in the shape of Daniel Guggenheim, the wealthy industrialist. Goddard obtained funds to set up his experiments in Roswell, in the desert of New Mexico. But after a few good years, during which one of his projectiles reached a height of 3 km, higher than any rocket had previously flown, Goddard suffered a couple of disasters. In 1938 high winds blew down his launch platform. Then two years later one of his rockets exploded on the launch pad. At around this time there was talk of Goddard teaming up with other American engineers to mount a joint rocketry programme. Among those interested were workers at the California Institute of Technology and the newly formed National Advisory Council for Aeronautics. But Goddard refused to get involved, saying obstinately that he could do better if he stuck by himself.

It was not that Goddard failed to try for government support. He thought his ideas would appeal to the military who, so he reasoned, could use rockets as weapons. In 1941, a year after the USA entered the Second World War, Goddard sent the War Department a memo about his work. It was ignored. Goddard did have a final swan song, though. In 1942, he worked at a naval establishment in Annapolis, Maryland, on liquid-fuelled rockets for helping aircraft to take off from carriers. However, true to form, ill fortune intervened. One of Goddard's devices exploded prematurely and this dissuaded his supervisors from taking his ideas any further. The inventor died in Baltimore in 1945, bitterly disappointed that no one in authority would take him seriously.

Goddard left a legacy of no fewer than 214 patents, roughly three for every year of his life. Many of these laid the foundations for the modern rocket business. Among the innovations for which Goddard was responsible is the notion of the multistage rocket by which an object is accelerated to a high velocity by a series of boosters, each of which starts firing after the previous one is finished. Today's space rockets are normally three- or even four-stage affairs. A single-stage vehicle would not provide enough power to take heavy satellites and other payloads into orbit.

This American rocket enthusiast also pioneered the use of gyroscopes to control vehicles in flight. A gyroscope senses the direction of movement of a vehicle and applies corrections to the propulsion mechanism to keep the craft on a set course. Goddard also studied the technique of 'gimballing' – mounting the tail sections of rockets on steerable joints to keep the craft upright during the initial stages of a launch. If it were not for gimballing, modern rockets would keel over as they tried to climb near-vertically through the lower part of the atmosphere just after leaving the launch pad.

Only in 1960 did the American government officially recognize Goddard's true worth. It formally admitted it had infringed the inventor's patents during the rocket-building programmes of the 1950s, and as compensation, federal officials handed over $1 million to Goddard's widow. The National Aeronautics and Space Administration, the body formed in 1958 to co-ordinate the USA's civilian activities in space, was sufficiently sensitive of the country's debt as to name one of its top space bases after him. This is the Goddard Space Flight Center near Washington DC.

If American government leaders were slow to realize the potential of rockets, the same could not be said for their counterparts in Germany. There the Nazi leaders of the 1930s and 1940s funded the devices as weapons of war and, indirectly, gave an impetus to the USA and Soviet space programmes that followed a couple of decades later. Paradoxically, as history was to show, the German effort was not quite vigorous enough. Had Germany pushed just a little harder, its rockets would probably have won the Second World War for Hitler, changing enormously the course of world events.

Work on rockets in the USA and Germany was linked in a singular episode in the summer of 1922. Robert Goddard, still working his way to his first set of rocket successes, received a letter asking for advice on rocket design. The writer was a young German student named Hermann Oberth. Oberth was a member of a group of German scientists passionately interested in space travel. Oberth's theoretical brilliance in rocketry was matched by his extreme politeness. He wrote to Goddard:

> Already many years I work at the problem to pass over the atmosphere of our Earth by means of a rocket. When I was now publishing the result of my examinations and calculations, I learned by the newspaper that I am not alone in my inquiries and that you, dear Sir, have already done much important works at the sphere. In spite of my efforts, I did not succeed in getting your books about this object. Therefore, I beg you, dear Sir, to let them have me . . .

In spite of the less than perfect English, we must assume that Goddard, being a kindly soul, sent the young German the information he required. Helped no doubt by this as well as other sources, Oberth made outstanding progress. In 1923, he wrote a far-seeing book, *The Rocket into Interplanetary Space*. This started with the following four bold statements:

1. The present state of science and technological knowledge permits the building of machines that can rise beyond the limits of the Earth's atmosphere.
2. After further development, these machines will attain such velocities

that they will not fall back to Earth; they will also be able to leave the zone of terrestrial attraction.

3. Such machines will be built in such a way that they will be able to carry men.

4. Under certain conditions, the manufacture of such machines might be profitable. Such conditions might develop in a few decades.

Oberth did not have to wait long before one of his countrymen tested out his ideas. In 1928, Frederich Stamer became the first person to fly in a rocket-driven aircraft. Stamer's vehicle travelled a little more than a kilometre in a trip of seventy seconds. He wrote afterwards: 'Flight with jet propulsion I consider exceedingly pleasant. Motor vibration and torque are absent; so that one has the sensation of pure gliding, and is reminded of the rockets only by their loud hissing.'

This was still the rocketry of the monied amateur who found experimenting with rocket engines fashionable and faintly amusing. But more threatening developments were under way. In 1929, the German army started looking at the military potential of rockets. In the area of armaments the Treaty of Versailles that marked the end of the First World War had severely restricted the German military forces. Anxious that the defeated country would not become a force in weaponry in the foreseeable future, the victorious allies stipulated in the treaty strict limits on the artillery units that Germany could deploy. But the treaty said nothing about rockets, to which military chiefs in Germany turned with eagerness.

The army initially tested rockets at a base in Kummersdorf, near Berlin. Trials took place with rockets powered by both liquid and solid fuels. Steady progress was made, despite a few accidents; for example, an explosion on a testing pad killed three people in 1934. In 1937, the army moved its testing establishment to Peenemünde, a town on the Baltic coast that offered more space. By 1942, three years after the start of the Second World War, the base employed 6,000 people. It was adjacent to another research centre run by the air force. This was charged with developing a pilotless aircraft, the V1, or the 'doodle bug', as it was known to Londoners who were on the receiving end of the device during the latter stages of the war.

At the army's Peenemünde base, Germany's top technicians worked on the rockets that, so they thought, could make an important contribution to military victory. The base managed to launch small rockets, code-named A2, A3 and A5, and then produced the A4: a monster 14 metres tall that weighed 13 tonnes, one tonne of which (during operational flight) could be high explosive. The A4 became better known as the V2, the first modern military rocket and the first true prototype of later space vehicles. The designation 'V' for both the rocket and the unmanned aeroplane was applied by the German Ministry of Propaganda in 1944. The V stands for 'Vergeltungswaffe', meaning 'revenge weapon'.

During 1942 two attempts were made at launching the V2. Both failed. The war was starting to swing away from Germany. Then, on 3 October, the first successful flight went ahead. Major General Walter Dornberger, the commander at Peenemünde, recorded in his book *V-2* (published in 1954) how events turned out. He wrote:

> It was an unforgettable sight. In the full glare of the sunlight, the rocket rose higher and higher. The flame starting from the stern was almost as long as the rocket itself. The fiery jet of gas was clear-cut and self-contained.

He went on:

> My heart was beating wildly. The experiment had succeeded. For the first time in the history of the rocket we had flown an automatically guided, jet-driven projectile as far as the limit of the atmosphere . . . and sent it on into practically airless space. We had been working ten years for this day.

Dornberger was exaggerating just a little. The V2 did not quite reach the top of the atmosphere, not, at any rate, by today's definition of where this is. It attained, none the less, an altitude of some 80 km and travelled 200 km. Nothing made by man had previously travelled further from the Earth. The previous record was held by a shell fired by an artillery vehicle of the First World War, the so-called Paris Gun. As Dornberger put it in a speech to his research workers the evening after the triumph: 'To land, sea and air may now be added infinite space as a medium of future intercontinental traffic.'

For the Germans in 1942, however, more immediate problems were at hand. Launched from its bases in Germany, so the rocket's proponents calculated, the V2 could hit targets in the big cities of Britain, spreading havoc and possibly knocking the country out of the war. Britain's intelligence services were slow to respond to the threat, and only in August 1943 did the RAF bomb Peenemünde, killing 735 people. Most of these, unfortunately from the UK's point of view, were low-skilled technicians, leaving the key team of rocket designers relatively intact.

The mistakes were not all on Britain's side. Hitler was by no means speedy in backing his rocket designers, but eventually, the month before the bombing raid, he put his country's full weight behind the project, ordering the Ministry of Munitions to turn out the rockets in factories. The chief plant was at Nordhausen, in what is now East Germany, where 8,000 workers, many from concentration camps, laboured in appalling conditions to make up to 900 of the devices per month.

From September 1944 to the end of the war in the summer of the following year, the Germans launched only about 3,000 V2s, and many of these failed to reach their targets, mainly the cities of London or Antwerp. Yet the V programme was hardly ineffectual. Some seven times as many V1s as V2s lifted off the ground. Between them the weapons killed 13,000 people and injured twice this number. In Greater London alone, they destroyed 23,000 homes and damaged 100,000 more.

The V devices were hugely expensive. They cost Germany some $3 billion, at 1940s prices ($20 billion–$30 billion at 1983 prices), according to the estimates of Frederick Ordway and Mitchell Sharpe in their book *The Rocket Team*, an authoritative book on rocket development in the 1940s and 1950s. Yet if only Germany had put more resources behind the weapon at an earlier stage of the war, the outcome could have been different. As Major General Dornberger put it: 'The use of the V2 may be aptly summed up in the two words "too late". Lack of foresight in high places and failure to understand the technical background were to blame.'

For Dornberger and many of the other Peenemünde engineers, however, the Allied victory was not an end but a beginning. In May 1945, when the war was stuttering to a close, the American army, by then cutting a swathe through Europe, captured an important group of Peenemünde workers. The key catch was Wernher von Braun, the head of the V2 design team. The capture took place in Oberjoch, in south Germany, to which many of the Peenemünde staff had fled, fearing the take-over of their research base by the Soviet armies which were pressing through the northern part of Germany. (Peenemünde was overrun by Soviet troops on 5 May 1945.)

It was at this point that the Yankee skill at striking a commercial bargain shone through. Realizing the worth of the prize, American government officials made a deal with von Braun and his men. The price of their freedom was that they came to the USA (on lucrative salaries) to work on rockets for the American taxpayer. This was how von Braun, Dornberger and other prominent German engineers built the rocket that, in 1969, would take to the Moon men who had been their bitter enemies two and a half decades earlier.

Others among the victorious Allies were also eager to claim the spoils of war. Like the USA, the Soviet Union and, rather less enthusiastically, France and Britain regarded the Peenemünde engineers as a package of concentrated brainpower, ownership of which might yield considerable benefits. The Germans were not impartial about whom they worked for after hostilities ended. In the words of one engineer at the time, quoted in *The Rocket Team*: 'We despise the French; we are mortally afraid of the Soviets; we do not believe the British can afford us; so that leaves the Americans.'

The American drive to import German know-how and equipment took the cryptic name Operation Paperclip. In total, the USA gave jobs to 500 Peenemünde engineers and shipped across the Atlantic the parts for 100 V2 rockets. In 1946, the defectors settled down at the US Army's White Sands proving ground in New Mexico. This was only about 200 km from Roswell, the place where Goddard had conducted his (largely unheralded) experiments in the 1930s.

Between 1946 and 1952, engineers launched sixty-four vehicles based on the V2, mainly for scientific experiments. For example, they took into the upper portions of the atmosphere instruments that measured quantities such as temperature and radiation levels. The adapted rockets were more powerful than the V2s of wartime and commonly reached heights of 180 km, above the limits of the atmosphere. Then on 24 February 1949 came another milestone. The world's first two-stage rocket was fired into the heavens from the proving ground. More powerful than anything that engineers had devised hitherto, the rocket reached a record-breaking altitude of 400 km. Undoubtedly, the German émigrés, who formed a key part of the US rocket team, used designs sketched out at Peenemünde as part of what had been called the A9/A10 project. This had been on the drawing board in 1941–2: a two-stage vehicle based on A4 technology which, had it ever been built, would have had a range of 5,000 km, enough to reach eastern America from a base in, say, Portugal.

The British were not totally left out of the scramble to claim something from the German rocket work. In October 1945, the German town of Cuxhaven was home to a series of bizarre launchings of the V2. Rather than being on the receiving end, the British were in charge. Engineers tried out the rockets, in what was called Operation Backfire, to glean information about how the vehicles worked. Later, the government brought twenty Peenemünde engineers to work in Britain, but the research petered out owing to lack of funds.

In one extraordinary episode, the Americans tricked their British comrades-in-arms out of a cache of important blueprints on the workings of the V2. The German rocket team under von Braun had buried the documents in a mine shaft in Dörnten. US officials were first to the mine, telling their British counterparts, who were also on the scent, that they were collecting samples of iron ore.

The USSR was also keen to salvage something from the engineering efforts of the Germans. By reaching Peenemünde itself before the other allies, and then occupying other establishments such as the Nordhausen rocket factory (which it took over from the Americans in July 1945), the Soviet Union had a good chance to tap the expertise of the rocket designers. After the war, the Soviet government brought 2,000 of the engineers, sometimes forcibly, to rocket establishments in the eastern USSR.

Unlike the USA, the Soviets already had considerable experience in rocketry. Drawing on the work of establishments such as the Gas Dynamics Laboratory near Leningrad, Soviet troops used rockets during the war as antitank weapons. Also the country had its own rocket range at Kapustin Yar near Stalingrad. By October 1947, Soviet engineers had launched from there some twenty V2s which they had expropriated from the defeated Germans.

As a result of this ingrained expertise (and perhaps also because of a suspicion about working too closely with the Germans) the Soviet Union never assimilated its own rocket workers with the Peenemünde émigrés. Rather, the German team was kept as a kind of consultancy unit that worked on projects on the periphery of the main Soviet effort. According to some observers, only about 10 per cent of the Soviet advances in rockets during the 1950s were due to the efforts of the German workers. In the case of the United States, the comparable figure was probably around 80 per cent. Indeed, in 1960 the US Air Force worked out that without Operation Paperclip America would have had to pay out an extra $2 billion to reach the level of expertise it had attained in rocketry.

In both the Soviet Union and the USA, much of the rocket work after the war centred on military applications. As the cold war developed, the Allies grew further apart. They poured immense effort into building missiles – logical developments of the V2 – that could threaten a target up to several thousand kilometres away. The high explosives used in the V2 warhead were replaced by something more terrifying: the nuclear bomb. An advance in the technology of missiles in one country was matched by a similar advance in the other. Both sides protested all the time that they would not use the weaponry. Rather, it was to be held back as a threat in case the other side made an attack. Thus was born the peculiar strategy of nuclear deterrence which (at least so its advocates claim) to this day has kept the Soviet Union and the USA from fighting each other.

The train of developments was not quite the same for each country. The USA was better at making small bombs (in size, that is, not necessarily in power). This was partly because of American advances in miniaturization of electrical circuits. The Soviet Union, on the other hand, had to be content with big and heavy bombs. Thus they developed the more powerful launchers.

In the USA, von Braun and his fellow workers were integrated in 1950 into the army's Redstone Arsenal, established in Huntsville, Alabama, the previous year. In 1953, they produced the Redstone missile, a marginally more powerful version of the V2 that carried a nuclear bomb and had a range of 320 km.

Other components of the military were keen to get in on the missile-building act. The US Air Force formed a ballistic missile division in 1954, working on long-range (intercontinental) missiles such as the Thor and Atlas. The US Navy, meanwhile, won an informal contest between the different branches of the military forces to build a rocket that would do something other than deliver a warhead to an enemy power. In 1956 the Naval Research Laboratory started work on a vehicle called Vanguard. This was intended to put into space America's first satellite. The US government wanted to launch the satellite in International Geophysical Year, 1957–8. It was to demonstrate the purely peaceful role that space technology could play. But Vanguard was never more than a partial success. On the first attempt to launch it, in December 1957, the rocket blew up in front of millions of TV viewers. Of eleven efforts to launch scientific satellites with Vanguard, eight were failures.

The US Army designed the vehicle that launched the USA's first satellite into orbit. In 1956, the army had formed a Ballistic Missile Agency at Huntsville under the control of von Braun. The agency was later to be transferred to the civilian NASA and formed the kernel of the new organization's Marshall Space Flight Center. Von Braun's rocket – the Jupiter-C, based on Redstone technology – lifted the Explorer 1 satellite off the ground on 31 January 1958.

The sense of triumph of the American engineers was, however, not all it could have been. They knew they were only second best. Almost four months earlier, the Soviet Union had won the race to launch the first artificial satellite. On 4 October 1957, Soviet engineers enabled the 83 kg Sputnik 1 to be lifted into space on board a modified two-stage intercontinental missile. The craft orbited the Earth at an altitude of up to 900 km. It measured physical quantities such as the density of electrons in the upper atmosphere. The satellite's radio transmissions, as it passed over the world, signalled that people would never look up to the heavens in quite the same way again.

3 ☆ Communications from the Sky

Imagine you are far out into space, about a tenth of the way to the Moon (approximately 36,000 km from Earth). Our planet would appear a large orb set in the blackness of the sky; to be exact, it would look twenty times larger than a full Moon viewed from the Earth. There would be stars to gaze at and, naturally, the Earth's nearest neighbour itself, which would appear a much bigger and more significant body than the white blob in the heavens to which we have grown accustomed. Otherwise, though, life would be pretty lonely for any traveller who had reached this point after a journey equivalent to flying between London and Los Angeles five times over.

Lonely, that is, except for some strange and unexpected phenomena. Huge chunks of metal, some of them as tall as a building of several storeys, would occasionally drift into our astonished star traveller's gaze. The hunks of hardware would vary in shape. Some would be cylindrical, like giant dustbins. Others would be box-shaped, with long slivers of silvery material protruding from either side, like the wings of a giant albatross.

The space voyager would have stumbled into a region of space which, against most well-understood rules of logic, constitutes some of mankind's most precious real estate. Unlike most other places that the materialistic peoples of the world squabble over, this resource contains absolutely nothing. That is if, of course, you discount the minute quantities of gas molecules that may be present this far from the Earth's surface.

This segment of the heavens is a perfect position in which to place communications satellites. The enormous 'birds' encountered here are packages of integrated circuits, amplifiers, antennas and other assorted electrical equipment. The vehicles pick up radio signals transmitted from the Earth's surface, boost the message to a higher power, and then send the amplified signal to a different point on the globe far below. (Purists will forgive the use of the last word; in space, of course, there really is no such thing as 'below'.)

This location for satellites is known as the geostationary orbit. It constitutes a ring 36,000 km out into space above the Equator. Strictly speaking, the orbit is not so much a ring as a giant torus; it is like a vast, though very thin, tyre which girdles the Earth far out from the planet and has a cross-sectional diameter of just 150 km. Satellites placed anywhere in the 'solid' part of this imaginary tyre can be said to have geostationary status.

It so happens, for reasons that will be made clear later, that satellites placed this distance from the Earth move round it at the same angular rate as the planet itself revolves about its axis. Hence craft in this position appear fixed in relation to an observer on the Earth; that is, so long as the craft are above the Equator. If the orbit is inclined in relation to this line, then the satellite will move in a different plane from the Earth. To a person on the ground it will describe a figure of eight. Satellites that are 36,000 km above the Earth but not above the Equator are said to be in a geosynchronous orbit (Figure 1).

Unlike our natural satellite, the Moon, geostationary satellites never set or rise. People manning ground equipment that directs signals to the hardware are never in any doubt as to the satellite's position. This makes the job of sending and receiving relatively easy. This giant celestial hoop has a circumference of 265,000 km, ten times the circumference of the Earth. So is this one resource where there is plenty of room for everyone? The answer is 'no'. Places in this orbit are contested fiercely by virtually every country and by a host of business corporations. It is one place where virtually everybody wants to be.

The International Telecommunication Union (ITU), an agency of the United Nations in Geneva, employs a large number of staff to monitor who wants to put satellites into this particular orbit and who operates them there already. At the end of 1983 the geostationary ring was home to about 200 satellites of which roughly 120 are active, that is still transmitting and receiving (Table 1, p. 32). The rest are still in position in or near the geostationary ring – they are considered defunct either because parts of their hardware have worn out or because they have drifted too far out of position to be tracked

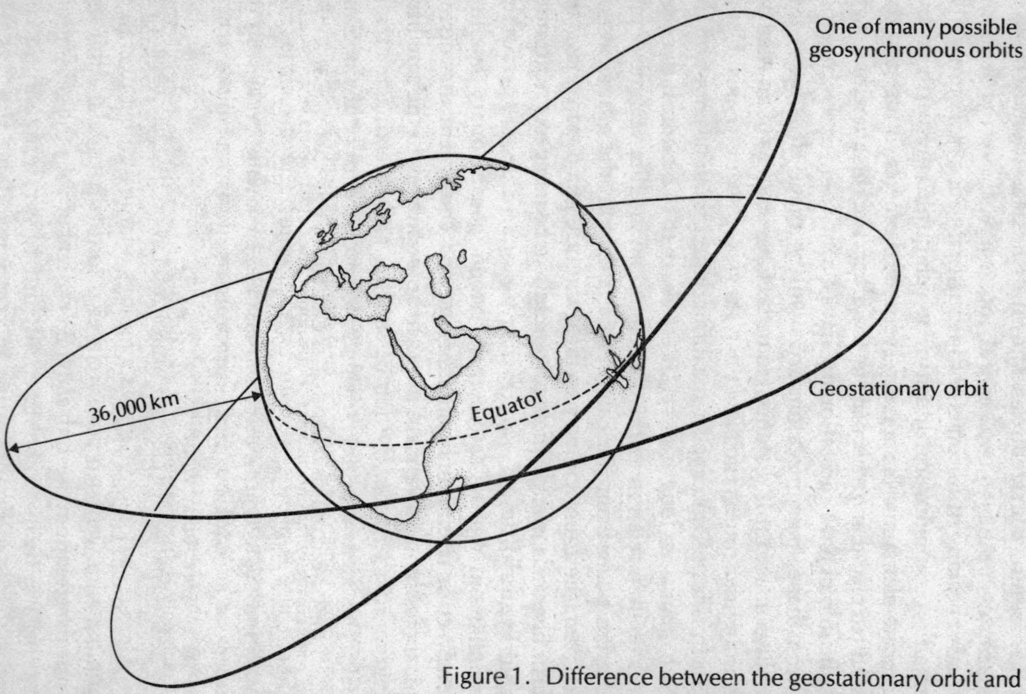

Figure 1. Difference between the geostationary orbit and geosynchronous orbits

from the Earth.* According to 1983 figures from the ITU, governments or private companies have applied for about 170 more 'slots' in the geostationary ring (these are in addition to the active satellites already in place). The parties concerned have reserved places for launches over the next five years or so.

To put matters in perspective, the geostationary satellites are just a small proportion of the 3,000 or so satellites that the world has launched since Sputnik 1 in 1957. Most of these, however, were put into relatively low orbits near to the Earth's atmosphere and, as a result of frictional forces, many have burned up. Space contains around 10,000 man-made objects, many of them parts of decayed satellites or rocket fragments, that are big enough to be tracked by ground-based equipment.

Slots for satellites in the geostationary ring are limited as a result of problems with radio interference. Owing to the demands of other users of the airwaves, the signals to and from geostationary craft must be packed into certain small sections of the frequency spectrum (see Table 4, p. 48). Insufficient frequency space exists for all the channels to have their own unique frequency slot, so, often, signals sent to different spacecraft have to share the same part of the spectrum.

In practice, operational geostationary craft (that is those receiving and transmitting signals) have to be separated by somewhere

* Nowadays the general rule is that satellites become inactive after five to ten years in orbit, though some spacecraft have shown greater longevity. Satellites stop operating normally for one of two reasons. Either their solar panels become faulty, thus cutting off supplies of electricity to power vital equipment, or the vehicles run out of the rocket fuel that powers their small position-keeping thrusters. As the satellites can no longer keep in the correct position for receiving or transmitting messages, they have no effective use.

Commonly, 'retired' satellites drift aimlessly about the geostationary belt. No one knows or cares very much where they are, which adds to the difficulty of compiling statistics regarding satellites whose useful lives are over. There is, however, a new trend whereby some satellite manufacturers are starting to ensure that their craft contain enough fuel near the end of their lives to be boosted into an orbit higher than the geostationary ring. This is a move designed to decrease the crowding of this orbit by satellites which are no longer serving any purpose. In an example of this exercise, ground controllers employed by the European Space Agency at the organization's satellite centre in Darmstadt, West Germany, in three manoeuvres on 24–25 January 1984, shifted a disused geostationary satellite called Geos-2 into an orbit about 300 km above the geostationary ring.

Table 1. Geostationary satellites in position to end of 1983[1] (not counting launch failures)[2]

Owner or category	*Total in position*	*Active*
Western communications craft		
Intelsat	28	17
US commercial[3]	22	21
NASA[4]	5	4
Other US[5]	5	3
Canada	8	5
Italy	1	1
Japan	5	5
European Space Agency[6]	3	3
France/Germany[7]	2	1
Indonesia	3	3
India	2	1
Subtotal	84	64
USSR communications		
Cosmos 1366/637	2	1
Raduga	13	13
Gorizont	8	7
Ekran	11	9
Molniya	1	0
Subtotal	35	30
Military communications		
Defense Satellite Communications System (US)	12	8
Initial Defense Satellite Communications System (US)	27	0
NATO	5	3
FLTSATCOM (US)	5	5
LES (US)	4	2
Skynet (UK)	2	1
Tacsat (US)	1	0
Subtotal	56	19
Military early warning (i.e. nuclear bomb detection)		
Rhyolite (US)	4	2

Owner or category	*Total in position*	*Active*
Integrated Missile Early Warning (US)	12	4
Early Warning Satellite (EWS) (US)	1	0
Cosmos 775 (USSR)	1	0
Subtotal	18	6
Meteorological and scientific		
Japan	1	1
Western Europe[8]	3	2
US[9]	8	4
Subtotal	12	7
Total	205	126

1 Figures based on International Telecommunication Union statistics and commercial sources. The figures may not include all military satellites, particularly those from USSR, and the distinction between satellites which are 'active' as opposed to 'defunct' is sometimes open to dispute as often there is a narrow dividing line between the two classifications.

2 Figures do not include failure of satellite on or soon after taking up geostationary position, nor launch failure due to rocket malfunction. Notable failures of these two types include: Insat 1A (India, April 1982); RCA Satcom 3 (US December 1979); Skynet 1B (UK, August 1980); Skynet 2A (UK, January 1974); DSCS 5 and 6 (US, May 1975); DSCS 9 and 10 (US, March 1978); Sirio 2 (Italy, September 1982); Marecs B (ESA, October 1982). Launch failures in 1984 include Westar 6 (US) and Palapa B2 (Indonesia) in February 1984; Intelsat V (F-9) in June 1984.

3 Western Union, 5 satellites (4 active); RCA Satcom, 7 satellites (all active); Satellite Business Systems, 3 satellites (all active); Comsat, 4 satellites (all active); AT&T, 1 Telstar satellite (active); Hughes, 2 Galaxy satellites (active).

4 4 Advanced Technology Satellites launched and 1 Tracking and Data Relay Satellite.

5 These include 3 Marisat maritime communications satellites (all active) and 2 Syncom satellites (both defunct).

6 Includes Orbital Test Satellite; Marecs-1; European Communications Satellite-1.

7 Symphonie satellites.

8 Includes 2 meteosats and 1 Geodynamics Experimental Ocean Satellite (GEOS).

9 Includes 6 Geostationary Operational Environmental Satellites (GOES) and 2 Synchronous Meteorological Satellites.

between two and four degrees of the geostationary arc, putting an obvious limitation on the number of vehicles that can be parked in the ring. This separation is sufficient to ensure that adjacent spacecraft that are receiving messages on the same frequency do not accidentally tune in to the wrong signal sent from Earth in a wide radio beam. The minimum permissible separation for operational craft works out at about 1,500 km – greater than the distance between Warsaw and London. In fact, space vehicles in the geostationary orbit are often closer than this because of the large number of inactive craft. There is, however, relatively little danger of spacecraft colliding. Statisticians have calculated that the odds of one geostationary vehicle bumping into another, as a result, for example, of a fault in a propulsion system, are of the order of 30,000 to 1.

Not all the hardware in the geostationary ring is for communications (Table 1), but despite the other applications for satellites, which account for about 15 per cent of the vehicles in the ring, the huge interest in the geostationary orbit stems overwhelmingly from mankind's incessant enthusiasm for transferring information. This statement needs some qualification. Communications satellites are just one part of the process that started when our caveman ancestors reasoned that, with pieces of wood, they could scratch rudimentary symbols in the dirt. These crude implements were the first tools in what we now call information technology. Other hardware followed, gradually becoming more sophisticated: pen, printing press, typewriter, radio, telephone, television, computers and satellites. All are mechanisms that transfer sounds, words, pictures and symbols which have some bearing on human thoughts or knowledge.

Transferring information was important at the time of Stonehenge; the 'information age' has been around a long time. None the less, the business of shifting this particular commodity from place to place has become today perhaps the most dominant factor affecting commercial and social life.

The telephone shows how devices that help in this information transfer have mushroomed. The humble handset is one of this century's most popular scientific inventions. In 1920, just 25 million people had telephones. The figure grew to 50 million in 1945; 175 million in 1965; today the world contains some 500 million telephones and it estimated that this figure will double by the early 1990s. Telephones are devices for transferring the information content of the human voice. By themselves they are not much use. Telephones need a mechanism to link them. The best established methods use land lines such as copper wires. Such wires can also pass between land masses in the form of submarine cable. Satellites form another way of linking such devices. Signals are sent from large exchanges to an orbiting craft and then back again to Earth. Satellite transmission proves useful when messages have to be sent between two centres widely separated by the curvature of the Earth.

In the early days of communications satellites, the messages sent on this journey into space almost entirely comprised telephone traffic. For example, satellites made (relatively) cheap telephone calls possible between people in the USA and Europe. The satellites also carried TV pictures and sound, and the phrase 'live by satellite' entered the language as satellites relayed pictures of events thousands of kilometres away to people's living rooms.

Today, however, the satellite channels are filled not just with human talk and pictures but with computer messages. The vehicles provide the links between a variety of computerized devices. If you listened in to one of these messages, you would not understand very much. You would hear a barrage of blips – the 0s and 1s of the binary code which constitutes the language of computers. Sending such information from one place to another via satellite is becoming increasingly important.

The messages can encompass huge quantities of business or financial information shunted between computer banks. The satellites can also transfer signals with which, for example, a computerized control console operates a machine tool in a factory where the console and the tool could be the width of a continent apart. For instance, Boeing, the US aircraft manufacturer, runs machines in its Seattle factories with signals sent from a computer room on the east coast near Washington DC. Mitsubishi, the Japanese engineering company, has used a similar satellite connection to send blueprints of aeroplane parts from a design office in Texas to the company's aircraft factory in Nagoya, Japan.

Military messages are also important. As Table 1 shows, many of the geostationary craft are for defence, with the satellites linking up armed forces in different parts of the globe. They also send computerized signals to memories in weapons such as missiles (see Chapter 6).

Communications craft can relay huge quantities of computer data. Modern space vehicles act as a conduit in which one ground station sends to another up to 1,500 million binary digits of information each second. At this rate the satellite would take about sixteen hours to transmit the entire 20 million books in the British Library.

What of the nature of satellites in general? Satellites are artificial objects set in motion about the Earth. Rocket power provides sufficient energy to boost the hardware against the force of gravity to a convenient place in space. Rockets also supply the satellite with the angular velocity needed to keep it in position. A simple analogy shows how satellites stay in orbit. Consider an object such as a tennis ball that, fixed to a piece of string, can be kept in motion by whirling it above the head. The circular motion of the ball results in an outward thrust called the centrifugal force. Opposing it is a centripetal force. This tries to draw the ball towards the centre of the circle and is channelled along the string. So long as the string is taut and the ball revolves at sufficient speed, the forces are in equilibrium and the ball continues to rotate. The equilibrium may, however, be upset. The ball could move too fast; then the outward force may be too great and the string would break, causing the ball to fly off at a tangent. Alternatively, the ball may slow down; the string slackens and the ball falls.

An object rotating about the Earth follows similar principles. The difference is that the centripetal force transmitted through the string is replaced by the natural, unseen force of gravity. For equilibrium to be maintained, the satellite must move at a speed sufficient to provide a centrifugal thrust that balances the force of gravity. Gravity decreases with distance, so the higher above the Earth a satellite travels, the slower it will need to move in order to remain in orbit.

How does a satellite get into orbit in the first place, without falling back to the Earth? Imagine that a huge tower about 500 km high sits on the Earth. The top is therefore in free space, beyond the effects of 'drag' caused by the air molecules in the atmosphere. Objects are fired from this tower at varying speeds (Figure 2). The slowest

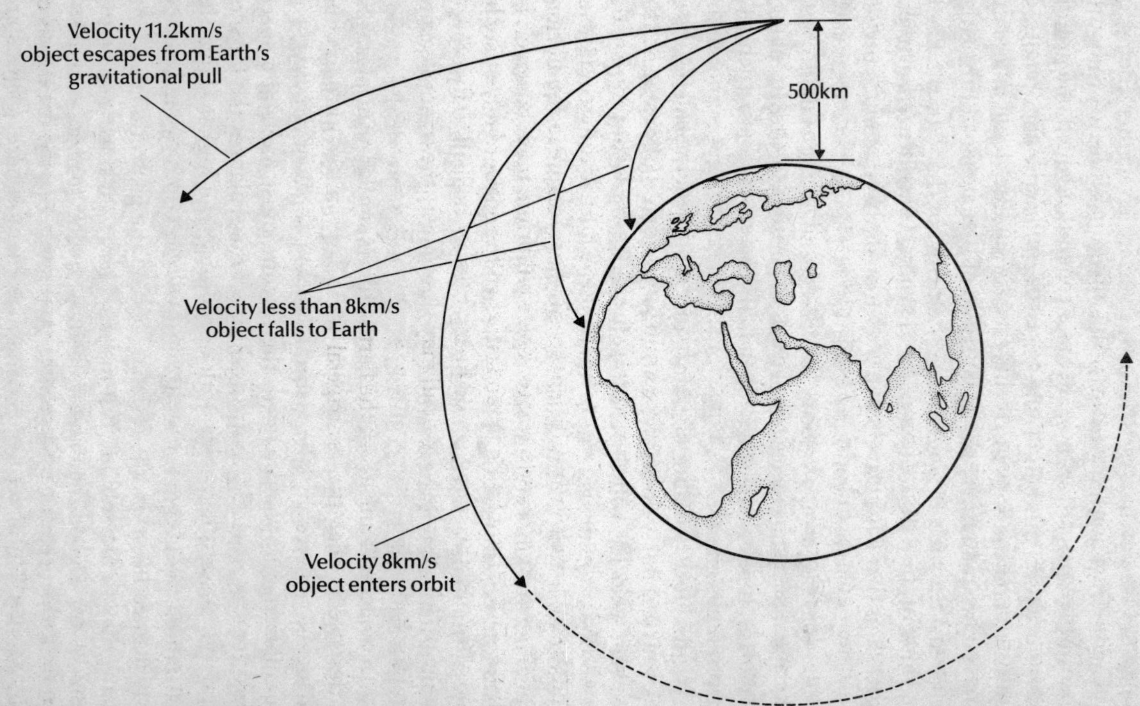

Figure 2. Theory of how a satellite enters orbit

objects do not travel very far before falling back to Earth, but an object ejected at a very high speed travels so fast that by the time it starts to 'fall', it misses the Earth's surface altogether. Being in space, there is nothing to slow down the piece of hardware. It continues to circle the planet at a height of 500 km; in other words, it enters orbit. At this height, the velocity at which the object 'misses' the Earth and continues in a circle is about 8 km/s (29,000 km/h); this is more than thirty times the speed of a jet airliner. If the body reached 11.2 km/s it would leave the neighbourhood of the Earth and head off into the solar system. This is the velocity needed by the spaceships that have travelled to the Moon or the unmanned probes that have explored deep into space. Other planets have their own values for this 'escape velocity' which depend on the planets' mass and diameter. Thus tiny Mercury has an escape velocity of 4.3 km/s while for Jupiter, the giant planet, the escape velocity is nearly fifteen times greater.

If, therefore, satellite operators can propel their vehicle to the relevant orbital velocity, it will stay in orbit indefinitely. The relevant velocity will depend on the satellite's height above the Earth. Put another way, the further above the planet the satellite is, the longer it will take to make a complete revolution. A vehicle travelling at 8 km/s 500 km above the surface would sweep round in about ninety minutes. This is the orbital period of relatively low-altitude spacecraft such as the USA's space shuttle. At the very great distance from the Earth of 380,000 km, the orbital period is twenty-eight days. This, of course, is the distance of the Moon, which travels through space relatively slowly, at about 1 km/s (3,500 km/h), to prevent itself from spiralling to the Earth. It follows that a particular orbit exists in which the time to make a complete revolution is exactly one day, the same as the Earth's period of rotation. This happens to be at 36,000 km, which explains why the geostationary orbit has its peculiar characteristics. A geostationary satellite travels at a little over 3 km/s (Figure 3).

Explained like this, the rationale behind the geostationary orbit seems easy to understand. But mankind had been staring up at the heavens for a good many years before anyone realized the application of the orbit 36,000 km high. Arthur C. Clarke, the scientist and science fiction writer, produced the first proposals for geostationary satellites in the form of a memorandum to the British

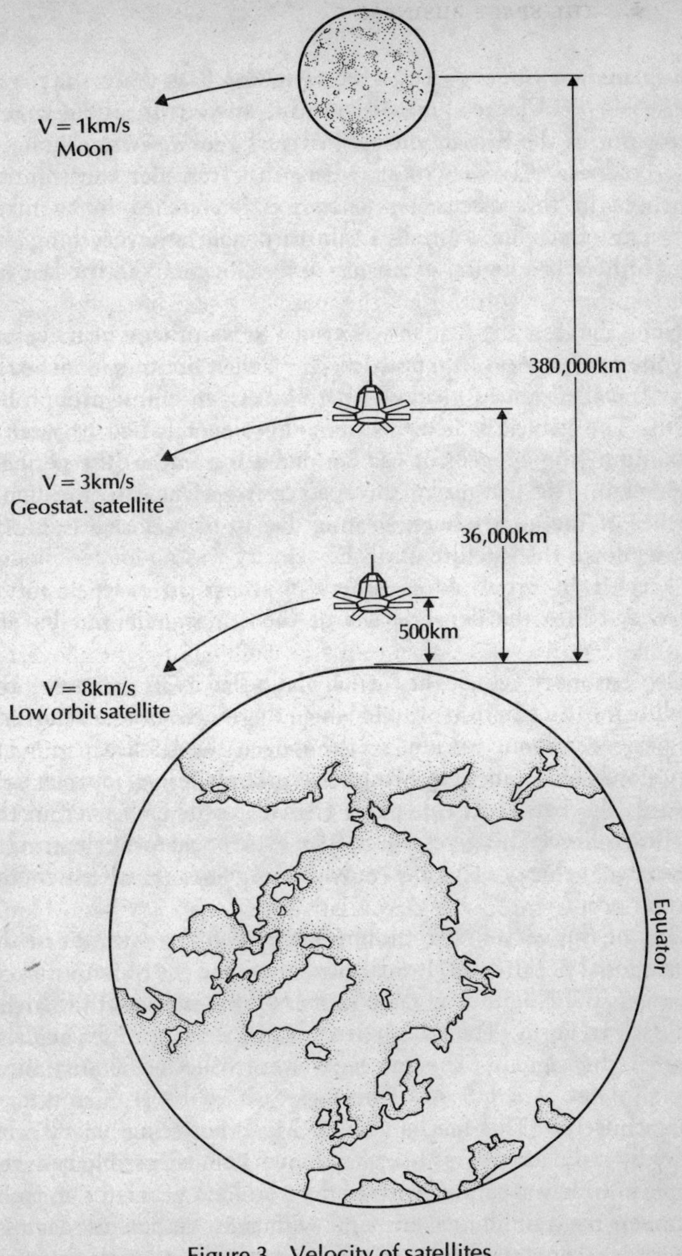

Figure 3. Velocity of satellites

Interplanetary Society. The document, dated 25 May 1945, was the basis of Clarke's article on the subject in the October 1945 issue of the British journal *Wireless World*. With characteristic boldness, Clarke wrote: 'Many may consider the solution proposed in this discussion to be too far-fetched to be taken seriously. Such an attitude is unreasonable as everything envisaged here is a logical extension of developments in the last ten years.'

None the less, the British writer had to wait many years before anyone put his ideas into practice. The Soviet Sputnik launched in 1957 travelled not in a circular path but in an ellipse around the Earth. The vehicle's distance from the planet varied between a maximum (the apogee) of 941 km and a minimum (the perigee) of 227 km. The perigee of the spacecraft was near to the upper reaches of the atmosphere, so drag due to air molecules quickly slowed down the satellite until the velocity was no longer enough to keep it in orbit. After ninety-two days in space, Sputnik descended into the denser parts of the atmosphere and burned up.

Geostationary vehicles are usually launched by first injecting the satellite into an elliptical orbit in which the apogee is 36,000 km and the perigee is about 300 km. At the apogee, the satellite is moving very slowly and is about to swing back again on its long journey back towards the Earth. At this point a small motor embedded in the satellite is fired. This gives the craft an extra boost to accelerate it to the correct velocity of 3 km/s, at which stage the vehicle enters the circular orbit.

One of the very first communications vehicles was not in the geostationary orbit at all. It was none other than the Moon. In 1959, engineers tried 'bouncing' radio signals off the surface of the Moon and back to Earth. The radio telescope at Jodrell Bank in England emitted the signals. The messages were collected almost three seconds later at a US Air Force research centre in Cambridge, Massachusetts. (This mechanism of 'Moon bouncing' is still used today by radio amateurs. It makes conversations possible between people in widely separated parts of the world.) A year later, in 1960, engineers tried similar experiments with another 'passive' device – one which contains no hardware to boost or otherwise process signals – which this time was man-made. Echo-1 was simply a balloon 30 metres in diameter which reflected radio waves sent between California and New Jersey.

The world's very first communications satellite was a vehicle called Score, which US engineers launched in December 1958. It carried electrical hardware to amplify the radio signals sent from the ground. This vehicle relayed a taped Christmas message from President Eisenhower. Courier, a similar experimental satellite placed in an elliptical orbit, followed in October 1960. It was developed by the US Army Signal Corps, which wanted new methods for keeping infantrymen in contact with headquarters. Neither Score nor Courier transmitted messages instantaneously, however. They accepted a signal, stored it on a tape and then relayed the message to its destination some time afterwards.

The world had to wait until 10 July 1962 for the first communications satellite that sent on messages immediately, or, in engineering jargon, 'in real time'. The vehicle, Telstar, was also the first satellite to be paid for by a commercial organization. AT&T, the American telecommunications company, spent $50 million on the craft, which is best remembered today for the catchy tune to which it lent its name.

With Telstar, AT&T relayed telephone calls and TV pictures between Europe and North America. It could handle, at any time, sixty telephone calls or one TV signal, but there were several snags. The main one was its orbit, which was highly elliptical. The satellite sped round the Earth on a path that varied between 5,600 km and 950 km from the surface. Thus, from any point on the globe, the 'rising' and 'setting' times varied from day to day and so Telstar connected ground stations in Britain with the USA for only a period of twenty minutes on most days. This was the time when it was in the right position over the Atlantic Ocean. That was all very well for short, 'showpiece' TV relays, of which there were several, or for people who did not mind booking up in advance for telephone calls at a specific time. But the arrangement hardly made for an efficient commercial service.

Another problem was that the satellite relayed signals of a very low power, about 2 watts. Imagine trying to see, from a distance of thousands of kilometres, a light bulb shining with one-fiftieth of the intensity of the lamp in your living room. Then you can understand the difficulties of picking up signals from Telstar. In practice, it meant using very large dishes, tens of metres in diameter, the sort with which astronomers collect weak radio signals transmitted from far-away galaxies.

Finally, the state of the art at that time in solar cells was not very advanced. Telstar, like the satellites of today, needed solar cells to convert radiation from the Sun into electricity. This was for powering the craft's amplifiers and other electrical equipment. Telstar's solar devices wore out quickly. After four months in orbit, the vehicle no longer responded to signals.

Three scientists at the Hughes Aircraft Company in the USA, with little doubt the world's most experienced satellite builders, are credited with converting the theoretical concepts for a geostationary vehicle into reality. The scientists were Harold Rosen, Donald Williams and Tom Hudspeth. They worked on a series of three Syncom satellites for NASA in the early 1960s. The first vehicle failed. In July 1963, Syncom 2 entered a geosynchronous (as opposed to geostationary) orbit above the Atlantic. In August 1964, Syncom 3 became the first geostationary vehicle. It hovered above the Pacific and relayed TV broadcasts of the Tokyo Olympics to the rest of the world.

Hughes followed this up with the first commercial communications satellite in the geostationary ring. It went into orbit above the Atlantic in April 1965 and entered full service two months later. This was Intelsat 1 (Early Bird), which was operated by an unusual body called the International Satellite Communications Organization, known simply as Intelsat. Intelsat provides an impressive example of co-operation on a global scale. The organization is based in Washington DC. Shares in it are owned by 109 nations. The leading countries are the USA, UK and France, which have stakes of 24, 13 and 6 per cent respectively. The shareholdings include small fry such as Burkina Faso, Uruguay, Tunisia and the Vatican City, which each own 0.05 per cent.

The USSR and its communist allies have never joined Intelsat. They prefer to run their own satellite system for communications called Intersputnik. This leases Gorizont satellites that the USSR operates mainly for the benefit of its own national communications system. Besides its Gorizont satellites, the Soviet Union also has a series of Raduga and Ekran craft in the geostationary orbit. Ekran vehicles are mainly for relaying TV. The USSR also has some vehicles in elliptical orbits that are for communications, for instance the Molniya series.*

Intelsat is the world leader in launching satellites. To June 1984, it had successfully put into orbit twenty-nine craft (seven launch attempts failed), worth nearly $1 billion (Table 2). At any time, the organization operates a total of around seventeen satellites, some of which are purely back-up craft in case the main ones go wrong. The vehicles are above the Atlantic, Pacific and Indian oceans. As one geostationary craft can 'see' about one third of the globe, in this way Intelsat provides links between virtually every non-communist country. Some 400 ground stations in 170 nations or territories are linked to the Intelsat system.

International telephone calls are growing annually by about 15 per cent in most developed countries. In 1978, Intelsat had enough capacity to channel 12,000 telephone calls simultaneously; today the number has more than trebled. Calling up people long distance is one of the few pastimes that is becoming cheaper. The annual rent for a full circuit of Intelsat satellite capacity – what you would need for a telephone conversation – fell from $64,000 in 1965 to $9,200 in 1983. Table 3 shows how international telephone calls have become less expensive since the days when the only available option was using submarine cables.

Intelsat 1 resembles today's satellites only in the way that a First World War triplane bears a family likeness to Concorde. Intelsat's first vehicle was a cylinder half a metre high weighing 39 kg. It looked like a large can with a few antennas sticking out. With two transponders, devices for receiving and retransmitting radio waves, it could relay a total of 240 telephone calls. The satellite radiated 40 watts. The sides of the can were covered with solar cells to produce

* Molniya satellites are generally in an orbit that varies between 40,000 km above the northern hemisphere and 500 km above the southern hemisphere. One Molniya satellite is therefore above the Soviet Union for some eight hours at a time. A full period of communication is ensured by using three such satellites that take over the service in succession. The Gorizont satellites are the most advanced of the USSR's family of geostationary craft. The first was launched in 1978. One Gorizont satellite, above the Atlantic, handles international communications between the Soviet Union and European countries, together with Cuba and Algeria. A second vehicle, above the Indian Ocean, channels messages to and from Asian countries. The Ekran and Raduga satellites are used solely for internal communications. One of their uses is to beam TV transmissions around the country. (Further details are given in Table 1.)

Table 2. Intelsat satellites

Generation	*1*	*2*	*3*	*4*	*4A*	*5*	*6*
Year of first launch	1965	1967	1968	1971	1975	1981	1986
No. launched[1] to geostat. orbit	1	3	5	7	5	8[2]	—[3]
No. of transponders	2	1	2	12	20	29	46
Power (watts)	40	75	120	400	500	1,200	2,180
Mass in orbit (kg)	39	86	152	732	863	950	1,700
No. of voice channels	480	480	2,400	9,000	12,000	25,000	40,000[4]
Frequency band[5]	C	C	C	C	C	C&Ku	C&Ku

1 Does not include launch failures.
2 To 30 June 1984; seven more Intelsat 5 satellites planned.
3 Five Intelsat 6 vehicles planned; with options on eleven more depending on demand.
4 70,000 voice channels possible with time-division multiplexing technique. Ku band: 11–14 GHz.
5 C band: 4–6 GHz.

Table 3. Cost (1983 prices) of a three-minute telephone call between the UK and other countries

	1961	*1983 (standard rate)*
Fiji	£22.50	£3.71
Australia	£18.00	£3.71
USA	£18.00	£1.88
Poland	£5.50	£1.24
W. Germany	£3.00	£1.24

electricity from the Sun's radiation. The craft was 'spin stabilized': it rotated continuously like a spinning top, the movement keeping the vehicle in roughly the right position in orbit.

By contrast, the current generation of Intelsat satellites, the Intelsat 5 series, are each 6 metres tall and weigh almost 2 tonnes (Figure 4). They have a power output of 1.2 kW and can handle 12,000 simultaneous telephone calls, using no fewer than twenty-nine transponders. The capacity of the next generation, Intelsat 6, which is being built ready for launch in the late 1980s, will increase yet again to 35,000 telephone calls.

The shape of the satellites in the geostationary orbit is gradually changing. The Intelsat 5 series are like giant birds. They have a pair of solar panels that splay out from a central capsule containing the transponders and electrical equipment. With the panels fully deployed (they are tucked away into the side of the central capsule during the launch, ready to be unfolded once the vehicle is safely in position), an Intelsat 5 satellite is 16 metres long. It appears that vehicles of this shape will eventually replace the dustbin-type craft like Intelsat 1. Intelsat has, however, reversed this trend by choosing a set of 'can' satellites for the Intelsat 6 series. The 'bird' satellites have to keep in the correct orientation in orbit, so that their antennas point in the right direction. This is done through small rocket thrusters set in the body of the spacecraft. Under commands from ground stations, these fire small bursts to tilt or turn the vehicle so that it retains the proper aspect in space.

Satellite operators are increasingly favouring the bird satellites for two main reasons. First, to keep the craft in the correct position in orbit is more straightforward with this design. Thus the bird craft

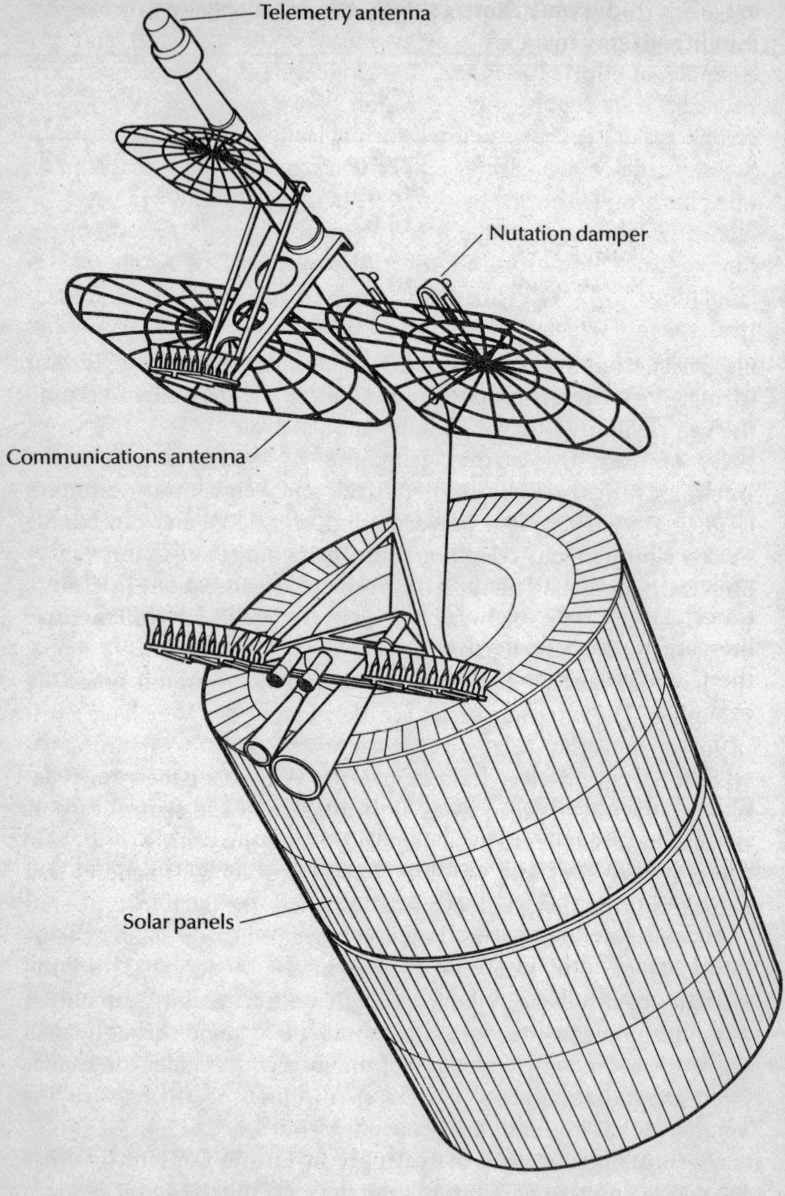

Figure 4. Intelsat VI satellite

are likely to become more popular as the requirement grows for satellites to stay in a tightly defined place in the sky, as a result, for example, of efforts to squeeze more vehicles into the geostationary ring and thus reduce the separation between adjacent craft. The second reason is that satellites with solar panels stretched out on 'wings' can provide greater amounts of power to the communications hardware which they carry. This is because the wings can be elongated (although an important restriction is whether or not they can be packed tightly into the rocket or space shuttle that is launching the satellite) in order to receive large volumes of radiation from the Sun which can then be converted to electricity. As satellite operators are requiring ever more powerful satellites, the winged design seems likely to gain the upper hand. With the can design, in contrast, room for solar cells is limited to the cylindrical surface of the satellite itself.

The most important limitation with which satellite operators have to contend is that regarding frequency space. Radio waves can travel only in a certain range of frequencies, and much of this radio spectrum has been allocated to uses other than satellite transmissions. In practice, operators of communications satellites are normally forced to use frequencies in what are called the C and Ku bands (Table 4), C-band satellites being the more common.

In C-band satellites, the message from the ground station to the satellite is squeezed into a 500 MHz 'spread' of frequencies around 6 GHz. The signal from the satellite to the ground uses the same amount of frequency space but this time around 4 GHz. In the Ku band, the two key frequencies are 14 and 11 GHz for the uplink and downlink respectively.

Today's satellite operators, therefore, have 500 MHz of what is called 'bandwidth' in which to cram their radio signals. In the satellite communications business there used to be a convenient rule of thumb. You needed a bandwidth of 40 KHz to carry the sound of a human voice – half a telephone circuit. You needed about 1,000 times as much frequency space, 40 MHz, to carry a high-definition TV picture. Thus, within the full 500 MHz bandwidth, a satellite operator could arrange to have twelve identical transponders on his space vehicle, each carrying 1,000 voice channels or one full TV picture (this calculation leaves 20 MHz of the full bandwidth left

Table 4. Frequency bands for satellite transmission

Band	*Frequency (GHz)*	*Remarks*
P	0.3–1	Used by scientific/ Earth observation satellites
L	1–2	
S	2–4	
C	4–8	Commonly used for communications satellites
X	8–12.5	Reserved for military spacecraft
Ku	12.5–18	Increasingly used for new communications satellites
K	18–26.5	
Ka	26.5–40	High frequencies under investigation for future use
V	c. 50	High frequencies under investigation for future use

over for contingencies). So a satellite working to these principles could carry a total of about 6,000 telephone calls. Changes in technology have made this rule of thumb outmoded for modern satellites, but it still works for some of the older vehicles. For example, the Intelsat 4 series followed the rules fairly well. Seven Intelsat 4 spacecraft were launched between 1971 and 1975, and two were still operating in 1983. Each had twelve transponders and were able to carry 4,000 telephone calls plus two TV pictures.

In recent years, engineers have squeezed more information-carrying signals into the same amount of frequency space. They have done this in two ways. The first uses the technique of polarization of radio waves. It is possible to arrange for the electric field associated with radio waves to be in either the horizontal or vertical plane. Thus two messages can be sent with the same frequency if the two signals are polarized differently in this way. So twice as many messages can be crammed into the same amount of frequency space.

In the second method of increasing information-carrying capacity, engineers have modified the coding techniques in which the messages are sent. Traditionally, this coding is done by modulating a carrier wave in terms of changes in frequency or amplitude.

In newer techniques, the information is coded into a digital format – 'blips' of 0s or 1s. With such methods, far more information can be squeezed into a given frequency space. This is why, by sending 'blips' through space rather than conventional waveforms, with the new Intelsat 6 satellites engineers think they may transmit as many as 100,000 telephone calls at any one time. Plans also exist for a still more advanced Intelsat 8 satellite, to enter orbit in 1997. It would handle 240,000 telephone conversations, or an equivalent amount of data or TV transmissions.

With polarization and new data coding techniques, and by increasing the size and power of satellites to pack more communications equipment on board a single satellite, the number of transponders can be increased. Thus, while current communications satellites generally have between twelve and twenty-four transponders (of which only perhaps a third may be switched on at any one time), future satellites may have two, three or four times as many. Ford Aerospace, for instance, plans a new generation of spacecraft with fifty-four transponders. Hughes, meanwhile, has designed a satellite with forty-eight transponders. One transponder, to use another rough rule, can transmit between 50 and 120 million binary digits of computer information per second. A satellite with fifty-four transponders would, therefore, have the capacity to channel some 6,500 million bits of data each second between two Earth stations. As forty bits are needed to characterize the average-length English word, a space vehicle of this kind would be able to transmit each second the information content of about 2,500 books the length of the one you are reading.

Over the next few years, the problems for satellite operators will be reduced in two other ways. First, engineers will simply find more frequency space for satellite transmissions. A precedent for this is that the first communications satellites used only the C band at 4–6 GHz. Gradually, the Ku band was explored at higher frequencies, and the first few craft to use this band are now in orbit. In general, the higher the frequency, the more difficult it becomes to design and operate transmission and receiving equipment. Now higher bands are being examined in the 20–30 GHz region which, at the moment, are not used for communication. Several satellites on the drawing board will feature experiments in these frequencies, including the European Space Agency's L-Sat, and Postsat and Athos, two satellites designed by West Germany and France.

A second option for satellite operators is to place their hardware closer together in the geostationary orbit. This will alleviate overcrowding, and provide more room in this orbit for the nations which have not yet staked a share in this important ring in space. In one method, engineers could load onto big 'antenna farms' a series of transponders that would today have to be placed on perhaps as many as a dozen individual satellites. In one study in 1983, Intelsat researchers proposed a satellite for the early twenty-first century which would measure some 240 metres square. The 50-tonne satellite would carry up to 5 million two-way voice circuits simultaneously.

To ensure that beams transmitting on the same frequency do not overlap and cause interference, engineers are working on 'shaped' beams. These will direct signals to small areas instead of, as is usual in most communications satellites, regions of the world that could be as large as a continent. With shaped beams, a satellite can direct one set of messages to country A while sending another lot of signals to adjoining country B. Even though in both cases the frequency band would be the same, no confusion would be caused because the beams would not overlap.

Another rule of thumb about satellites is that, as frequencies become higher and as the power that the craft direct into shaped beams increases, the size and complexity of receiving antennas are reduced. To receive signals from the first geostationary satellites, people needed an antenna costing hundreds of thousands of pounds. The antennas were as big as bungalows and contained huge volumes of electrical equipment. These are the kind of receiving dishes that shot to prominence in places such as Goonhilly in England and Pleumeur-Bodou in France.

Nowadays, people can tune in to signals from some geostationary vehicles with receivers that are small enough to place on roof-tops and cost only a few hundred pounds. The implications of this in TV reception will be discussed in the next chapter. Advantages arise for companies where the business of transferring information is important. Such organizations are installing small satellite aerials on their own premises. With these, computer information can be transferred or telephone calls made to other offices up to a couple of thousand kilometres away. This trend is most evident in the USA, where several companies offer business services involving satellites. One of the firms, Satellite Business Systems (which is partly owned by IBM), operates several satellites in the geostationary orbit which are expressly built for use in this fashion by big business corporations.

According to projections, the skies above North America will become chock-a-block with satellites over the next decade. On 1 January 1985, twenty-three active vehicles owned by US companies were in operation, plus several Canadian craft. If planned launchings go ahead, the total will come to more than forty by 1987. US companies have applied to the US government to orbit some eighty new satellites, some of them to replace existing ones, between 1984 and 1990 (see Tables 5, 6A and 6B). According to one study by

Table 5. Orbital allocation plan for US/Canadian/Mexican satellites. Plan released by US Federal Communications Commission, 27 April 1983 (applies to geostationary satellites to be launched to 1987)

Orbital position (degrees West longitude)	*Operator or satellite name*	*C/Ku bands*
143	RCA Satcom 5	C
141	Unassigned	C
139	Satcom 1R	C
137	Unassigned	C
134	Galaxy 1	C
132	Rainbow	Ku
131	RCA Satcom 3R	C
130	ABC 1	Ku
128	American Satellite	C/Ku
126	RCA	Ku
125	Telstar/Comstar	C
124	SBS	Ku
122	Spacenet	C/Ku
120	USSS	Ku
119.5	Westar 5	C
117.5	Canada	Ku
116.5	Mexico	C/Ku
113.5	Mexico	C/Ku

Orbital position (degrees West longitude)	*Operator or satellite name*	*C/Ku bands*
112.5	Canada	Ku
111.5	Canada	C
110	Canada	Ku
108	Canada	C
107.5	Canada	Ku
105	G Star	Ku
104.5	Canada	C
103	G Star	Ku
101	Unassigned	C/Ku
99	SBS	Ku
98.5	Westar 4	C
97	SBS	Ku
96	Telstar	C
95	SBS	Ku
93.5	Galaxy 3	C
93	Unassigned	Ku
91	Spacenet 3	C/Ku
89	SBS	Ku
88.5	Telstar	C
87	RCA	Ku
86	Westar	C
85	USSS	Ku
83.5	RCA Satcom 4	C
83	ABC	Ku
81	American Satellite	C/Ku
79	Rainbow	Ku
78.5	Westar	C
77	RCA	Ku
76	Telstar	C
75	Unassigned	Ku
74	Galaxy 2	C
73	Unassigned	Ku
72	RCA Satcom	C
71	Unassigned	Ku
69	Spacenet 2	C/Ku
67	RCA Satcom	C

USSS, US Satellite Systems Inc.; ABC, Advanced Business Communications Inc.; Telstar satellites operated by AT&T; Spacenet and G Star satellites operated by GTE.

Table 6A. US domestic satellite launches (to end of 1984)

1974	Westar[1] 1 (D)
	Westar 2
1975	RCA Satcom 1 (D)
1976	Comstar[2] 1 (D)
	Comstar 2
	RCA Satcom 2
1977	—
1978	Comstar 3
1979	Westar 3
	(RCA Satcom 3)[3]
1980	SBS 1
1981	SBS 2
	Comstar 4
	RCA Satcom 3R
1982	Westar 4
	RCA Satcom 4
	RCA Satcom 5
	Westar 5
	SBS 3
1983	RCA Satcom 1R
	Galaxy[4] 1
	Telstar[5] 301
	Galaxy 2
	RCA Satcom 2R
1984	Spacenet[6] 1
	Telstar 302
	SBS 4
	Spacenet 2

1 Owned by Western Union
2 Owned by Comsat, leased to AT&T
3 Satellite disappeared after launch
4 Owned by Hughes
5 Owned by AT&T
6 Owned by GTE
D satellite defunct
R replacement satellite
Satellites (active) in orbit at end of 1984, total 23: 4 Westar; 6 RCA; 4 SBS; 3 Comstar; 2 Galaxy; 2 Telstar; 2 Spacenet

Table 6B. US communications satellites due to be launched (to 1990) (This shows launch applications to US Federal Communications Commission to 11 January 1984. Not all launches have been approved by FCC.)

1984
Westar 6 (launch failure)
SBS 4 (successful launch)
Galaxy 3 (launch postponed)
G Star 1 (launch postponed)
G Star 2 (launch postponed)
Spacenet 1 (successful launch)
Spacenet 2 (successful launch)
Telstar 302 (successful launch)

1985
Westar 7
RCA Satcom Ku-1
RCA Satcom Ku-2
G Star 3
Spacenet 3
ASC 1
Telstar 303
Cablesat 1/Cablesat 2

1986
RCA Satcom 6
SBS 5
Galaxy 4
Spacenet 4
ASC 2
USSS 1
USSS 2
Rainbow 1
Rainbow 2
ABC 1
ABC 2

1987
RCA Satcom Ku-3
SBS 6
SBS 7
Galaxy 5
Galaxy 6
ASC 3
Colsat 1
Colsat 2
Equatorial 1
Equatorial 2
NE K1
NE K2
NE C1
NE C2
Rainbow 3
Rainbow 4
Ford 1
Ford 2

1988
Comstar Ku-1
Comstar Ku-2
SBS 8
Galaxy 7
Hughes K1
USSS 3
Federal Express 1
Federal Express 2
Digital Telesat 1A
Martin Marietta 1
Martin Marietta 2
NE K3
Testar 304
Westar 8
Westar A
Westar B
Ford 3

1989
Hughes K2
ASC 4
USSS 4

Table 6B *continued*

Digital Telesat 1B	*1990*
NE K4	Westar 10
Westar 9	Satcom H2
Westar C	SBS 9
RCA Satcom H1	
Comstar K3	
Aurora 2	

Westar operated by Western Union; Galaxy by Hughes; G Star and Spacenet by GTE; Telstar by AT&T; Comstar by Comsat, Colsat by Columbia Communications; Cablesat operated by Cablesat General; Aurora operated by Alascom. ASC, American Satellite Company; SBS, Satellite Business Systems; USSS, US Satellite Systems; Equatorial, Equatorial Communications; NE, National Exchange.

Western Union and ITT, the number of satellite transponders serving the USA will increase to 1,280 by the year 1990, compared to 680 in 1980. (The study assumes that one transponder handles a bandwidth of 30 MHz and bases its calculations on this.)

The figures show that the stakes in the satellite world, at least in the USA, are becoming higher. Companies must pay up to $100,000 to put a satellite in orbit. This includes constructing the craft and then launching it with a vehicle such as the USA's space shuttle or the Ariane rocket developed by Western Europe. For a full satellite system, a company must normally buy three separate craft; one is operational; one is a 'spare' in orbit and another stays on the ground as a back-up ready for launch if the other two fail.

As in any business, the US satellite industry is subject to peaks and troughs of demand. With the boom in satellites in orbit – and with the large number of applications to place more such vehicles in space – have come fears that insufficient demand could exist for all the transponders that are due to be inserted into the heavens over the next few years. If, for instance, the market for transmissions of TV from space grows at less than the projected rate (see next chapter), then many owners of American satellites could have large numbers of spare transponders. In early 1985, there were signs that such a glut in satellite hardware could be appearing. Whereas a couple of years before, satellite owners were renting out channels on their equipment (to organizations such as communications and TV companies) for about $2 million a year, reports in the industry put the 1985 price at about a fifth of this figure, indicating a steep fall in demand.

There is no doubt, however, that the satellite habit is spreading away from the shores of the USA. Western Europe plans a series of five publicly owned communications satellites, the first of which was launched in June 1983. These so-called European Communications Satellites, operated by a body called Eutelsat with headquarters in Paris, will channel telephone calls, TV programmes and business data to both large and small antennas. In Europe, privately owned corporations may start to operate satellite networks, following the trend established in the USA. Up to now the only vehicles for running such systems have been the state-owned postal and telecommunications administrations. In the UK, the private telecommunications firm Mercury and several merchant banks are interested in this proposal. Orion and International Satellite, two companies in the USA, are attempting to set up transatlantic satellite links. In November 1984, the US government announced it would permit companies to provide such services that would open up Intelsat, the satellite 'club' owned by state bodies, to competition with the private sector.

Outside Europe, Indonesia, Pakistan, South Korea, Japan, Argentina, Australia, Brazil and the Arab nations of the Middle East are among the countries which either operate or plan satellite networks. According to some estimates, up to 300 commercial communications satellites may be launched between the mid-1980s and the end of the century.

A booming area over the final years of the 1900s may be in satellite communications for vehicles such as cars, lorries, aeroplanes and ships. To send and receive signals from a fixed position is fairly easy; but little attention has yet been given to satellite links for people or things on the move. The exception to this is in shipping. An international organization called Inmarsat,* based in London, runs a satellite network for ships using its own spacecraft and others leased from bodies such as Intelsat. At the end of 1983,

* Inmarsat has about forty members. The leading countries are USA (23 per cent of the shares), USSR (14 per cent), UK (10 per cent), Norway (8 per cent), Japan (7 per cent), France (3 per cent).

some 2,000 ships had terminals for connection to the Inmarsat system. With these, people on board can make telephone calls or send data to other ships or to land bases. A ship's antenna must point at the right place in the sky. A ship is continually steering in different directions, in addition to rolling from side to side as a result of waves. Therefore the antenna must be steerable so it can set up a direct link with the spacecraft 36,000 km above the Earth in the geostationary orbit.

Similar receiving and transmission hardware may be deployed on aircraft to give airline crews, and passengers, similar facilities. Inmarsat is studying whether to start a communications service for aircraft using its own satellites. This may start up after 1988, when the organization's 'second-generation' spacecraft are due to be in orbit. Small antennas which could track satellites very accurately while an aeroplane is in flight would be required.

Another role for Inmarsat could be in operating a search and rescue service. This could be a successor to a satellite system called SARSAT/COSPAS that started in 1982. This enables satellites to 'listen out' for distress signals on a particular frequency, 121.5 MHz. These signals are triggered by emergency radio beacons. Some 200,000 ships and other vehicles around the world have installed these beacons which send a radio signal to the satellites when the vehicle has encountered an accident. The beacons can be triggered either automatically or manually. The radio messages are relayed to ground stations, which can work out the position of the stricken vehicle and guide emergency services to it. Three satellites have entered operation, two controlled by the Soviet Union and the other by America. France, Canada, Norway and Britain are other partners in the scheme. The system is also to be extended to a different frequency, 406 MHz. Ultimately a reliable worldwide service would require a great many satellites, in both low polar and geostationary orbits.

NASA's Goddard Space Center, which operates the US part of the emergency system, has reported that the SARSAT/COSPAS has saved 223 lives to June 1984. A recent rescue involved six fishermen in the Mediterranean who abandoned their vessel in high winds in February 1984. An aeroplane spotted them some 60 km from the place where their ship had sunk. Their lifeboat was within 7 km of the point in the sea that a mission control centre in

Toulouse, France, had predicted it would be – a testimony to the accuracy of the system. According to the latest records, the satellite network has saved 115 people from maritime emergencies and 107 in incidents in which aeroplanes force-landed. It has also located one person involved in a land emergency, a woman sled musher who became ill on an 800 km dog sled trip in Alaska in April 1984.

The difficulty of communicating by satellite with mobile vehicles is much greater on land than on sea. Ships can point their aerials at an orbiting satellite unencumbered by buildings and other obstructions. It is different for road vehicles, particularly in built-up areas where pathways between the vehicle and the satellite may continually be interrupted. Northerly countries such as Britain have particular problems. To receive signals from a spacecraft in a geostationary orbit, cars or other vehicles would have to point their antennas a long way south, to a pinprick in the sky over the Equator. The line of sight would make only a low angle with the horizontal. As a result, in towns the transmissions would be blocked by buildings.

Researchers at the UK's Rutherford Appleton Laboratory near Oxford have come up with an ingenious solution. This is to put a spacecraft in a highly elliptical orbit such that it hovers above Britain for roughly eight hours a day. During this period, vehicles would point their aerials directly upwards to set up a connection with the spacecraft.

The satellite could enter orbit in 1988, if government departments agree to put up the £20 million it would cost. The researchers say that the vehicle could help mobile radio users whose allocations in the existing radio bands are becoming increasingly crowded. The British spacecraft, or Communications Engineering Research Satellite to give the hardware its full name, would enter what is called a Molniya orbit, named after the Soviet spacecraft of this name (mentioned earlier) that are in similar trajectories. The British vehicle would reach its furthest point from the Earth over the UK, some 38,000 km into space. In this part of the orbit, the hardware would be travelling very slowly, which is why, in this position, it 'sees' the Earth for as long as eight hours. The British proposal could turn out a winner, if fully developed, with the concept being taken up by other countries. An operational system would require at least two satellites to provide continuous coverage.

Other countries are also talking about developing satellite systems for cars and lorries to provide telephone links. Canada and the USA may operate a joint satellite network, which uses a pair of satellites in the geostationary ring, by the end of the 1980s. These countries have plenty of wide-open spaces and, as a result, the problem of vehicles that attempt to 'see' the space vehicle but cannot, because of high buildings, would be less acute than in other nations where a higher proportion of the population is in built-up areas. Further, satellites, in general, prove economic when connecting points separated by long distances, which is why satellite systems for vehicles look immediately more promising in vast countries such as the USA and Canada.

In contrast, smaller countries, such as those in Europe, may find that a more economic answer to linking up cars and lorries to the public telephone network lies in new cellular radio systems. This hardware involves an amalgam of conventional radio and land-based telephone wires with which cars and lorries can transmit and receive messages via radio waves to and from small stations scattered around the country. Each station, in towns and cities, serves a 'cell' of a few kilometres square – the cells may be bigger in rural areas. The stations are connected by land lines to the ordinary telephone network. Computers in the radio stations automatically assign radio channels to vehicles as they move between cells. In this way, a driver can continue talking into his car telephone as he passes from one cell to another, oblivious of the fact that a different radio station is taking over the handling of his message.

The cellular radio systems in the USA (of which many are planned over the next few years, mainly in the big cities) may face stiff competition from the satellite network. The satellite system now under discussion follows from plans developed over several years by the Canadian government for a satellite called MSAT (the 'M' stands for 'mobile'). Originally, this was to be a non-commercial, experimental set of hardware. It was aimed at demonstrating how vehicles on the roads in Canada's vast rural areas could be linked to the public telephone network. More recently, commercial concerns have taken an interest in the plans. They think that, by the early 1990s, the satellites could provide communications for about 100,000 cars and lorries in the USA and Canada. The $400 million system would be based on two satellites in the geostationary orbit and several hundred ground stations. Each would serve road vehicles in a set rural area. Cars and lorries would need small aerials which, with their electronic parts, would cost about $4,000.

With the equipment, people would communicate voice or data messages from their vehicles to the nearest Earth station. From here the message would be channelled to one of the satellites and down again to another ground station, perhaps on the other side of the continent, which would connect up to the public telephone network.

Telesat, a Toronto company in which the Canadian government has a 50 per cent stake, would run the Canadian part of the service. Telesat already operates Canada's existing communications space vehicles. The US partner would be one of two companies – Skylink or Mobilsat. Both have applied to the Federal Communications Commission for permission to operate a service of this kind. Not only conventional telephone calls could be handled with the equipment. The satellites could act as relay points in the sky for data sent automatically from sensors on trucks or railway wagons. The sensors would provide information about the contents of the vehicles, for example the temperature of perishable goods such as food. By tuning into the signals from the satellite, haulage companies could keep a constant check on the goods carried by road and rail fleets. In another possible application, the system would send signals to transponders fitted to aircraft and pick up messages automatically emitted by the aeroplanes. In this way, the satellite hardware would identify aircraft and help to control air traffic.

Satellite networks of this kind will add more pressure to the congested geostationary ring. In recent years, the arguments over how to divide up the limited amount of room in this orbit have become more important. The whole issue will be discussed at two world gatherings organized by the International Telecommunication Union and due in 1985 and 1988. The meetings will attempt to define a set of rules for allocating slots in the geostationary ring to particular countries.

The most startling indication of the value placed on the geostationary ring came in 1976 at a meeting in Bogota, Colombia, of Third World nations, all of whom had in common that the Equator passes through their territories. The Bogota gathering declared that it had no faith in the way places in the geostationary ring are allocated by the International Frequency Registration Board, an arm of the International Telecommunication Union. This system, said the nations, favours the developed countries which have established places in the orbit before anyone else. It discriminates against the Third World, which may, in future decades, want to launch satellites into the ring and then find the orbit is clogged up.

The Bogota declaration said that any nation on the Equator automatically owns that section of the geostationary ring directly above it. Countries outside this small group of nations would be outlawed from operating spacecraft from this prime resource.

As a piece of diplomatic one-upmanship, the statement was hard to beat. Needless to say, the declaration had little practical effect – other than to cause lawyers to argue as to whether countries really can claim rights to outer space and, if they can, just where space legally begins.

The declaration did make plain, however, that the extension of mankind's influence to outer space would by no means always be accomplished in a spirit of co-operation. Bickering and argument would intervene far too often, and nowhere are the disagreements more evident than in the efforts to stake out a place in space for one of the twentieth century's most successful businesses – the TV industry.

4 ☆ Battle of the TV Beams

Few people have heard of the St Paul Rocks. They constitute a pinprick of land that protrudes from the storm-tossed waves of the Atlantic some 1,000 km off the coast of Brazil. The rocks are virtually on the Equator, at a longitude of 29.2° W. This barely visible speck on the maritime atlas is of very little value to anyone. Yet the rocks, or more specifically a portion of empty space above them, will soon be of great importance to people in Britain who watch television, which means almost everyone.

Later this decade a satellite parked in this position – to be accurate the craft will be over the Equator but a little further towards the South American coast at 31° W – will beam to Britain a new diet of TV programmes. According to current plans, the three-channel service will be operated by a consortium of the BBC, independent TV companies and non-broadcasting groups such as publishing houses and electronic firms. The craft, to enter orbit in 1988 or 1989, will be one of the world's first direct-broadcasting TV satellites and will represent one more addition to the crowded geostationary ring 36,000 km above the Equator.

To receive signals from the satellite, viewers will have to turn small dish antennas skywards so that they point towards the craft. This will be instead of tuning into TV programmes from ground transmitters with the conventional roof-top TV aerials. Britain is a long way north of the Equator. So householders who want good reception will have to ensure that their antennas, which will normally be less than 1 metre in diameter and cost a few hundred pounds, can point to an object rather low in the sky. Providing the elevation angle from the receiver up to the spacecraft is more than about 20°, and that the 'line of sight' is unobscured by objects such as buildings or trees, the householder should be able to receive signals from the satellite.

In theory, the best positions for the dishes will be in people's back gardens. However, particularly in towns, the view of the satellite from this point will often be blocked. This may lead do-it-yourself addicts to bolt the structures to their house roofs, but this could have unfortunate consequences, as many people's roofs are not quite up to bearing the weight of the heavy aerials. As a result, space-age TV could coincide with a boom for freelance builders as householders struggle to come to grips, sometimes literally, with a rash of broken joists and missing tiles.

Another practical problem for the TV viewer could be to distinguish between the signals from different satellites. By the 1990s, the nations of Western Europe will probably operate half a dozen or more TV satellites in the geostationary orbit. Each craft will beam programmes to the country which owns and operates it, but in practice the beams from the different satellites will overlap considerably. In some ways this is an advantage for householders who like watching TV, as they should be able to pick up broadcasts from many different countries. To do this, they will have to turn the antenna towards different satellites and tune the equipment to receive on the correct frequency. On the other hand, if the aerial is not pointing in the right direction, other signals may well be picked up unintentionally, which would lead to the crackly sound and fuzzy pictures of interference. The technical name for the overlapping of the beams or 'footprints' from broadcasting craft is 'overspill'.

On political grounds, the possibility of overspill has caused consternation among the governments of some countries who point out that satellite pictures from a neighbouring or distant country could spread propaganda to undermine cultural values, or even topple an unpopular administration. TV from space, on account of its greater powers of persuasion, represents a bigger threat than the beaming in of sound radio from other countries, which governments have used for propaganda since the Second World War.

According to many nations in the Third World and the Soviet bloc, governments should have the legal right to stop direct broadcasts from outside their countries if they do not like the programmes. Britain, the USA and other Western nations disagree. They say that such a right of veto would disrupt the principle of 'free flow of information', according to which as few barriers as possible should interfere with the passage of ideas.

The topic cropped up at an important debate at the United Nations in November 1982. On the vote to establish the right of veto, the Third World and the nations of Eastern Europe carried the day on a count of 88 to 15 with 11 abstentions. The vote was taken in a special political committee of the UN and was later upheld at the general assembly.

You do not have to be part of an oppressive or totalitarian regime to view with distaste some of the possibilities of direct broadcasting. Nelson de Jesus Parada, the director of the Institute for Space Research in Brazil, voiced his concern at the United Nations' conference on outer space (UNISPACE) in Vienna in August 1982. He said his country may hesitate to go in for satellite broadcasting as, once the nation establishes an infrastructure of households with the necessary aerials, they may be swamped by TV from other countries. Brazil's space supremo was clearly talking about the USA. In theory, American satellites would have no trouble beaming a diet of soap opera and baseball games into Brazil that not everyone in high office in Brazil's government would find appealing.

Friction over direct broadcasting has developed even between countries considered to be good neighbours. For years, the Canadian government has tried to restrict the power levels on which future TV satellites will broadcast. This is because, using high powers, it may be easy for American satellites to beam into Canadian homes programmes either of doubtful taste or which carry a high level of advertising centred on US-made goods. Authorities in the USSR, meanwhile, may be concerned over some elements of plans in the USA to begin direct-broadcasting services, probably in 1987. The American satellites will have to include Alaska in their 'footprints'. US satellite engineers acknowledge the danger that beams intended for this state could unintentionally spill over into eastern parts of the Soviet Union, bringing American TV to anyone there enterprising enough to install a satellite-receiving aerial.

Political arguments over direct broadcasting have also surfaced in Europe. The Compagnie Luxembourgeoise de Télévision (CLT), a mainly privately owned company in Luxembourg, wanted to launch a satellite to beam TV to a wide area of Europe, including France, West Germany, Belgium and the Netherlands. However, France raised objections. Government officials said they feared the service would beam English or American programmes which would do nothing to advance the cause of the French language. Also the French government, which has a strong hold over TV in France, would have no control over the nature of the programmes. CLT, the officials said disparagingly, was planning a 'Coca-Cola' TV channel.

Luckily for François Mitterrand, the French President, two of the major shareholders in CLT, the Hachette advertising agency and the bank, Parisbas, are owned by the French government. This, together with strong disapproval from Paris, led to the Luxembourg plan being quietly dropped in early 1983. A rather less grandiose scheme was announced in 1984 under which CLT is to rent two channels in the French TDF-1 direct-broadcasting satellite due to enter orbit in 1986. In this way, the French authorities will exert some control over the programmes carried by the Luxembourg company.

Something akin to the machinations over the plans of the tiny Grand Duchy will probably be seen in Eastern Europe. In particular, East and West Germany may well clash over TV from space. West Germany will probably have an operational direct broadcast satellite by the early 1990s. This would follow an experimental craft that the country will launch in 1986. The satellite will be in the geostationary arc at 19° W. By international agreement, any satellite that the East Germans launch would be a few degrees further along the geostationary arc, at 1° W.

The East German authorities would dislike their citizens tuning in to the West German craft, fearing the influence of Western propaganda. But how can people be prevented from receiving the signals illicitly? One factor would give the game away. As long as the householders' aerials point towards the East German satellite at 1° W, the authorities would have nothing to worry about, but they would view with suspicion any antenna turned to face other satellites, either the West German craft or, for that matter, a vehicle such as Sweden's, which is due to be launched around 1988. Perhaps this may lead to further policing in countries concerned about satellite broadcasts, with the orientation of roof-top satellite receivers being regularly checked.

Newspapers in East European countries have already voiced concern over the implications of direct broadcasting. *Horizont*, an East German government publication, illustrated an article on the subject with a communications satellite in the shape of a pistol with a white star on it and the word 'anti-Communism'.

Satellite TV is not always a cause for haggling between neighbouring countries. West Germany and France have shown exemplary co-operation. The two nations are jointly building satellites for their services, through a £140 million programme. France will probably be the first to offer a full direct-broadcast service, starting around 1987. The first French direct-broadcasting satellite, TDF-1, to be launched in 1986, will be followed later by a second satellite. Two satellites in orbit are required for a fully commercial service – the second craft acts as a 'spare' in case the first suffers a fault. Meanwhile, one of the main aims of the co-operation is that the two countries, through a joint company called Eurosatellite, plan to sell similar TV vehicles to other countries, for some £30 million–£40 million apiece. The main shareholders in Eurosatellite are MBB and Aérospatiale, the leading German and French aerospace companies.

The planning for TV satellites is done under the auspices of the International Telecommunication Union, an agency of the UN. The nations of Europe, Asia and Africa sorted out their allocations for TV satellites at a meeting in Geneva in 1977. North and South America had their own regulatory gathering in 1983. In both cases, the meetings ended with a neat plan indicating exactly where in the geostationary ring countries would be allowed to place their TV vehicles, together with a vast set of other regulations on, for example, the technical characteristics of the antennas needed to tune in to the broadcasts.

It is worth considering, just for a moment, that even twenty years ago people would have been astonished at the notion that countries from both the industrialized and the developing world would some day employ roomfuls of officials to worry about the niceties of such things as satellite orbits for weeks on end.

According to the plan for Europe and Africa, every country is allowed four or five channels of satellite TV, beamed from geostationary craft operating at frequencies between 11.7 and 12.5 GHz. This gives a total available bandwidth of 800 MHz. (The conventional telecommunications vehicles discussed in the last chapter normally send messages at the much lower frequency of 4 GHz, so there is no chance of radio interference between the two types of satellites.) Adjacent slots for the satellites serving Europe and Africa are separated by 6° of the geostationary arc to reduce the risks of interference.

Europe lies mainly between the longitudes 30° W and 30° E. Hence with 6° spacing between adjacent satellites in the geostationary ring, the number of positions is limited to ten. These positions also have to be shared by other countries on similar longitudes, in other words those in Africa. In practice, therefore, each slot is shared by up to a dozen nations. Each country's satellite will transmit on a slightly different frequency from within the allowable bandwidth. In Britain's case, the ITU conference grouped the country with ten other nations. The full list of countries with a slot for their TV satellite at 31° W is given in Table 7.

To give some meaning to the table, countries in groups 1 and 2 have to squeeze their signals onto the upper 400 MHz of the total available frequency range of 800 MHz. Groups 3 and 4 are allotted the lower bands, in other words the frequency space between 12.1 GHz and 12.5 GHz. Meanwhile, groups 1 and 3 polarize their signals in the opposite direction to the signals from satellites in groups 2 and 4. Overall, therefore, the total available bandwidth is split four ways. Within each 400 MHz portion of frequency space with signals polarized in a specific way, there is room for twenty channels of TV, each consuming 27 MHz (adjacent channels overlap slightly). Thus it is easily possible for three satellites to share this amount of space, as shown by the groups in Table 7. Each nation can broadcast five channels, leaving another five channels spare and claimed by no one. (The exception is group 1, where the frequency space is shared by only two satellites.)

Much of this is highly theoretical. Even in the case of a country such as Britain which has definite plans for satellite broadcasting, not all the channels will be taken up. Current plans are that only three channels will be used.* Countries such as the Ivory Coast or Liberia are a long way from even putting their plans for direct broadcasting down on paper; but if and when they get round to the idea, positions for their satellites will be marked out for them in space. It is the equivalent, in extraterrestrial terms, to putting your name down for a plot of land on which, some day, you may build a

* The actual frequencies on which Britain is allowed to transmit are 11.785, 11.862, 11.938, 12.015 and 12.092 GHz, according to the ITU regulations.

Table 7. Countries sharing TV satellite slot with the UK (31° W)

Group 1	*Group 2*	*Group 3*	*Group 4*
Sierra Leone	Spain	Britain	Cape Verde Islands
Burkina Faso	Ivory Coast	Liberia	Portugal
	Iceland	Ireland	Guinea Bassau

house. The whole exercise is most comforting to those sold on the idea of planning on a global scale.

Sometimes this in-depth planning can lead to anomalies. A few cropped up at the 1983 ITU meeting to sort out the positions for TV satellites to be operated by the North and South American nations. There was less difficulty here in thrashing out who would be entitled to what in the geostationary orbit. This is because of the geophysical fact that this part of the world contains fewer nations vying for a limited portion of the geostationary belt. Indeed, some countries came out of the conference with much more, in terms of satellite allocations, than they will ever need. For example, the tiny Caribbean island of Grenada (population 90,000) has the legal right to place in orbit no fewer than three satellites, beaming to Earth a total of twenty TV channels.

The Falkland Islands and associated British territories in the Antarctic region took up quite a bit of time during the 1983 conference. This was mainly because Argentina wanted to ensure that it is allowed to broadcast to and from the islands should Britain ever give up possession. Thus the conference approved the notion of a TV station in Port Stanley, the Falklands' capital, broadcasting, via a satellite, programmes to the Argentinian mainland. While the Falklands remain British, of course, it is most unlikely that anyone would want to do this – unless it is to beam to the South American continent pictures from Britain of expatriate Argentinian footballers playing in English sports stadia.

A more significant result of the deliberations in Geneva was that Britain obtained the legal right to launch two satellites to serve the Falklands and the Antarctic territories. The craft would beam to Earth a total of nine TV channels, in other words several more than TV viewers in Britain will receive from UK-operated satellite and terrestrial networks even when the country's direct-broadcast venture starts up. One of the Falklands' vehicles would be at 31°, over the famous St Paul Rocks, in the same position as the satellite that will serve the UK. By some judicious twiddling with antennas and other related electronics to ensure that the craft beams to the South Atlantic as well as Europe, the British satellite already on the drawing board could be converted to serve the Falklands as well. It adds up to an intriguing thought for anyone in Britain's Foreign Office with a little money at his disposal and wanting to cement ties between these lonely islands and the UK. There is a fair chance that the 'kelpers' on the Falklands would welcome satellite TV. For news from the outside world at present, they have to rely on a meagre diet of radio programmes from the BBC's external services unit.

These interesting possibilities notwithstanding, the most important task of the 1983 exercise in doling out satellite slots was to satisfy the demands of companies in the USA. This is the only country in the American region with firm plans to start satellite broadcasting. At the beginning of 1985, four firms had definite plans to start TV services from outer space – several others had filed applications to the US Federal Communications Commission asking for permits to begin satellite construction.* At the Geneva gathering, the USA gained permission to site satellites, owned by these firms, in up to eight slots in the geostationary orbit. Just for the record, these are at 61°, 101°, 109°, 118°, 147°, 156°, 165° and 174° W. Because the USA is so vast, a company seeking to broadcast across the whole of the nation would probably require two satellites, one for each half of the country. The space vehicles could share orbital positions with perhaps four or five other craft operated by rival companies, just as long as no two craft in the same slot broadcast on the same frequency.

Under the rules for the American region, TV satellites are allowed to broadcast in the range 12.2 GHz to 12.7 GHz, a total bandwidth of 500 MHz. In theory, this can be divided to provide up

* The companies with firm plans to start services were Dominion Video Satellite Corporation (due to begin broadcasts in 1987), US Satellite Broadcasting (1988) and Direct Broadcast Satellite Corporation (1989). Satellite Television, which is a subsidiary of Comsat, was originally to begin a satellite service in 1986, but the start-up date is now uncertain. Other companies that have announced an intention to enter satellite broadcasting are: Hughes Communications Galaxy, Satellite Syndicated Systems, National Christian Network, Advanced Communications Corporation, Satellite Development Trust and National Exchange Corporation.

to thirty-two channels of TV. Hence, with two satellites covering each time zone, each household with a suitable receiving aerial could tune in to around sixty or seventy channels of TV from space. This assumes, however, that companies which run the services take up the full orbital allocation. In practice, this is most unlikely to happen.

The broadcasting craft that the direct-broadcast firms will use provides us with a few interesting technical details about this new TV industry. Signals from TV satellites, as we have seen already, must be receivable by small dishes that can be fitted without much trouble to roof-tops or installed on odd scraps of land. In practice, this means packing a lot of power into the beam. Conventional telecommunications vehicles of the kind discussed in the last chapter beam signals in which each transmission channel has a power level of around 5 to 20 W. What is more, the beam is spread over a large area that could be the size of a whole continent. As a result, to collect the transmission some complicated electronics would be needed and an aerial large enough to pick up the very weak signals from outer space. With a TV satellite, the power level of each channel could be up to 200 W. Moreover, the beam itself is shaped so that it covers only a small area.

In theory, all the TV satellites that are eventually launched to cover the USA will beam into a carefully delineated area. In practice, however, a particularly well-tuned or sophisticated piece of receiving kit will be able to pick up signals that are not meant for it, perhaps from a space vehicle broadcasting into the adjacent time zone. This is again the problem of 'overspill' discussed earlier in the chapter.

The problem of 'overspill' crops up again in the United States – but in a slightly different form. Once more, people may pick up beams not intended for them. But the problem is not one of geography – no one will mind very much if citizens in, say, California tune in to broadcasts intended for people in Nevada – but of economics. People may pick up signals that they have not paid for, which could ruin the chances of TV companies earning enough revenue to cover costs. The obvious way round this is to scramble the broadcasts. Signals would be received only by viewers who have paid the relevant fee – a subscription of perhaps $20 to $30 per month – for decoding equipment fitted to their TV sets. The full cost of the receiving hardware (unscrambling equipment plus a dish aerial about 60 cm in diameter) would be $300 to $600. Nowadays there are some sophisticated methods of activating the decoding hardware that the apparatus would feature. For instance, once a subscriber has sent in his cheque for his monthly dose of satellite TV, a computer at the transmission centre would switch on the hardware by sending to it a special pulse of radio waves. The pulse would travel to the viewer's home in the same way as the TV signals themselves, that is by the satellite. But even this seemingly ingenious system can be beaten. With sufficient technological know-how, determined amateur electronics technicians (of which the USA contains a fair number) can decode signals not meant for them.

This kind of problem is not new for American TV companies. For years, they have tried to combat the people who tune in illicitly to the signals from space that feed cable networks. The USA contains some 5,000 cable stations, many of which receive their programmes from TV satellites via the kind of low-power telecommunications satellites considered in Chapter 3. These are the conventional kind that transmit in the 4 GHz region. (More modern craft are using the 12–14 GHz band.) The signals are received by a cable station with a big dish some several metres in diameter, before being fed into a cable network. In the USA, about 24 million households, roughly a third of all homes, receive cable television.

'Pirates', however, sometimes set up their own quite sophisticated aerials, costing perhaps $15,000, and tune in to the TV pictures from space for nothing. This saves them paying the subscription fee levied by the cable stations. It is also, by all accounts, a lot of fun for those who like to 'beat the system'. The habit of eavesdropping in this way to space signals is not confined to the USA. In Europe, electronics enthusiasts can tune in to twenty or more telecommunications satellites that also carry TV signals for feeding into ground-based networks. These include some of the Soviet and Intelsat craft (Tables 8A and 8B). According to estimates, the USA contains between 40,000 and 100,000 of these dedicated amateurs who spend their leisure hours picking up satellite pictures. One of them debated the legalities of the issue with British journalist Barry Fox:

> Look, if I hold out my hand and it rains, the rain falls on my hand. If I hold out my hand now, it's receiving satellite signals. The same signals are

Table 8A. Geostationary satellites observable from Britain. (Many of these satellites carry TV signals that can be picked up by radio enthusiasts with the correct equipment.)

Name	*Owner*	*Year of launch*	*Position* (degrees of longitude)
CS-2B	Japan	1983	77 E
Ekran 5	USSR	1980	72 E
Raduga 2	USSR	1976	72 E
Ekran 6	USSR	1980	64 E
Intelsat 5E	Intelsat	1982	63 E
Intelsat 5B	Intelsat	1981	60 E
BSE-1	Japan	1978	59 E
Intelsat 4AB	Intelsat	1976	57 E
Gorizont 5	USSR	1982	56 E
Ekran 8	USSR	1982	49 E
Raduga 6	USSR	1980	44 E
Raduga 9	USSR	1981	37 E
Raduga 11	USSR	1982	36 E
Gorizont 3	USSR	1979	13 E
ECS 1	ESA	1983	13 E
OTS 2	ESA	1978	5 E
Intelsat 4F	Intelsat	1974	0
Gorizont 7	USSR	1983	14 W
Gorizont 4	USSR	1980	14 W
Intelsat 4AC	Intelsat	1977	21 W
Intelsat 5C	Intelsat	1981	24 W
Raduga 7	USSR	1980	25 W
Intelsat 5D	Intelsat	1982	27 W
Intelsat 4AA	Intelsat	1975	31 W
Intelsat 5A	Intelsat	1980	34 W
Intelsat 4B	Intelsat	1971	39 W
Intelsat 4E	Intelsat	1973	42 W
Intelsat 4H	Intelsat	1974	53 W
Ekran 1	USSR	1976	55 W
Insat 1A	India	1982	57 W
RCA Satcom 2A	RCA	1983	67 W
TDRS 1	USA	1983	67 W
Galaxy 2	Hughes	1983	74 W
Westar 2	Western Union	1974	79 W

Satellites in position in September 1983. Some of the Intelsat satellites may be on the point of retirement. Source: N.L.H. Cresdee, *Satellite TV News*

Table 8B. Geostationary satellites observable from Britain – planned launchings

Name	*Owner*	*Year of launch*	*Position* (degrees of longitude)
Arabsat 2	Arab League	1985	26 E
Arabsat 1	Arab League	1984	19 E
ECS 2	ESA	1984	10 E
ECS 3	ESA	1985	7 E
Telecom 1A	France	1984	7 W
Olympus 1 (LSAT)	ESA	1987	19 W
Unisat 1A	Unisat (UK)	1988	31 W

Source: As Table 8A

falling on my house. Is that illegal? They are also falling on my 3-metre dish. Is that illegal? Of course not. So why shouldn't I do something with those signals that I never asked for in the first place? (*New Scientist*, 9 September 1982)

You might have thought that copyright legislation would say something about whether this electronic eavesdropping is legal, but the laws are out of date. They were drawn up in the 1930s, well before anyone thought of satellite transmissions, leaving the legal questions difficult to answer.

The pirating of signals in this way worries the organizations that plan satellite broadcasts. If a large body of people is already picking up signals from the sky for nothing, householders may think twice about having to pay to do the same thing. (Though, of course, the receiving hardware required for a 'true' direct broadcast service would be far cheaper and simpler to operate than that used by the electronic hobbyists who pick up signals at present.) The number of pirates is expected to grow. By 1986, the year before direct broadcasting is due to start in the USA, some 250,000 homes may be tuning in to TV in this underhand fashion. Aspiring satellite companies in Europe are faced with similar problems concerning how to safeguard themselves against piracy and infringements of copyright.

Organizations that want to run satellite TV services, in both the USA and Europe, face competition from the growth in activity in cable television. With more TV provided by wires passing people's homes, householders may not be prepared to pay out more money to receive yet more programmes from the heavens. The USA has seen more interest in cable than in Britain – at least up to now. In the USA, cable runs past perhaps two-thirds of the 80 million or so homes, though only about half these households bother to subscribe. The proportion of cabled households in Western European countries is much lower. The attraction to these nations of satellite broadcasting might, therefore, seem greater.

To complicate the picture, however, several countries in Europe are going all-out to cable up their urban areas, sometimes with a relatively new form of wiring, optical fibre. This has a much higher capacity to carry information than the traditional copper cable. Most significantly, it lends itself to piping large masses of computer data in the 'interactive' electronic services that may become popular in future. Such services include, for example, home banking and shopping in which people can arrange for cash transactions or deliveries of goods by keying in numbers to a computer terminal in their living rooms.

The growth of these networks has put a question mark against some of the more grandiose expectations for satellite TV. On the one hand, cable networks may take away some of the market for TV from space. Further, cable may become allied with certain kinds of satellites to damage the prospects for direct broadcasting. Low-power space vehicles are perfect for relaying to large antennas at cable 'head ends' programmes that are then fed by the ground-based wiring system into thousands of households.

Faced with the possibility that the alliance between low-power satellites and cable could threaten their plans, would-be direct broadcasters became a little less bullish towards the end of 1984. Certainly this was the case in Britain, where planners at the BBC decided against going it alone with their direct-broadcast scheme. Instead, they decided to spread the risk by operating in partnership with ITV companies and a group of non-broadcasting firms – Thorn EMI, Pearson Longman, Granada TV Rentals, Virgin and Consolidated Satellite Broadcasting. The start of the service is now more likely to be 1987 than the original date of 1986.

The main worry for the the UK's TV planners is that, to run a service, they face having to pay some £80 million a year for seven years to the Unisat consortium providing the high-power satellites (one operational and two spare). The consortium, comprising British Aerospace, GEC and British Telecom, is charging this high sum because of the difficulty and expense of building a high-power satellite. If the BBC has got its financial sums wrong, it will run the risk of being unable to recoup this cash from sales of its direct-broadcasting service.

According to some observers, the British TV companies may be able to operate a decent satellite service using a low-power satellite that broadcasts only about 20 W per channel. Such satellites – being already developed and in orbit – are much less expensive than the high-power variety. A test of this proposal was started in the USA in 1984. United Satellite Communications of New York, a company backed by the Prudential insurance company and General Instrument (the electronics firm), started a direct-broadcast service using channels rented from the Canadian Anik C2 satellite. The power broadcast per channel was only 17 W. But picture reception (using aerials slightly bigger than those required for high-power satellite reception) was by all accounts good. Unfortunately for United Satellite, the company stopped marketing its service in March 1984 after signing up for the service some 11,000 households in the Philadelphia–Washington DC area. This break came after the company ran out of money – it had by then spent $40 million – and by early 1985 it had still not restarted marketing operations, although satellite transmissions were still taking place. Whatever the fate of the company, the venture proved that TV broadcasts by low-power satellites can be perfectly viable. This will be still more so as the quality of domestic aerials improves, permitting them to pick up and convert to pictures relatively weak signals sent from space.

Waiting in the wings in the USA in the area of TV transmissions from the sky is News America Publishing, owned by Rupert Murdoch the newspaper magnate. At one time, the company was due to start a quasi direct broadcast using satellites owned by Satellite Business Systems. It seems now, however, that it will wait for the advent of the high-power satellites. Murdoch signalled his interest in this new entertainment medium by his purchase in 1983 of Satellite Television, a pioneering British company set up by an innovative ex-TV producer, Brian Haynes. Satellite Television is beaming programmes to cable head ends in Europe, using the ECS satellite, the first of a series planned by the European Space Agency.

In early 1984, the company for the first time channelled programmes to Britain, to a cable network in Wiltshire. Previously, TV viewers in Britain had been, strictly speaking, breaking the law if they picked up the company's transmissions. This is on the grounds that Satellite Television did not have the relevant licence from the Home Office. (Undoubtedly, however, British TV pirates have tuned in and risked the consequences.) By the end of 1984, the company was reaching in cable distribution some 2·8 million households in eleven European countries. At a later date, Satellite Television may well enter the area of direct broadcasting.

To discover the beginnings of this whole technology, we must go back to the mid-1970s. In 1978, in an experiment administered by Canada's Department of Communications, the Anik B spacecraft beamed TV pictures to several remote areas, including British Columbia, the Yukon and Northwest Territories. Japan has also launched several experimental direct-broadcast satellites, particularly to serve the country's outlying islands. Japan's first operational TV broadcasting craft, called BS-2A, entered orbit in February 1984, but failed only a month or so afterwards.

Similar ventures have been tried out in developing countries. The discussion in this chapter so far has been about the industrialized world. These countries already have TV, by courtesy of conventional ground transmitters that send signals to relatively small areas around them. Many nations in the Third World are lucky to have any TV at all. In many cases, their land areas are so vast that to install conventional terrestrial transmitters would be prohibitively expensive. Also, because of mountains or jungle, just erecting transmitters could be difficult. For these countries, the prospect of TV from space is most enticing. They would need only a satellite plus the necessary supporting hardware (and of course a good supply of programmes) to get TV off the ground at a fraction of the cost of the orthodox route.

India has been one developing country interested in the possibilities of TV from space. It has its own fledgling space industry which has been responsible for launching several satellites of its own. Under Prime Minister Indira Gandhi, the country's space programme received top priority. In 1975, India tried out a satellite TV service in an experiment with the US satellite ATS 6. The vehicle sent TV transmissions to some 2,400 villages in six 'clusters'. The programmes emphasized topics of an educational nature, for example series about public health. The Indian government thinks that satellite TV could greatly raise the consciousness of the population in areas such as this, so improving the people's standard of living. The country aims to start a full-time satellite TV service, broadcasting to selected villages which would set up small cheap receivers using India's own satellite, Insat-2.

China, too, could get into the satellite TV business. This is another country which has launched its own satellites, either for remote sensing or military surveillance. One application for the future that the country's ruling hierarchy is studying is for TV. At present only one-third of China's massive population of 1,000 million can tune in to conventional TV transmitted by ground networks. Sun Jiadong, the Vice-President of the Chinese Academy of Space Technology, told me in 1982 that more TV transmissions are needed desperately to raise the standard of education. A TV satellite could do the job. In 1984, the Chinese government started serious discussions with Western satellite firms about supply of the orbiting hardware. Jiadong believes that the job will be slow. He does not foresee his country starting a space-age TV service until the early 1990s at the earliest.

That, in all probability, is how satellite television will make its debut in most parts of the world – after a lot of rational discussion and perhaps after quite a bit of argument. Let us hope that the benefits – more varied TV for countries in the rich parts of the world and TV plain and simple for the developing countries that have never had the chance to get it before – outweigh the more nasty and threatening effects.

5 ☆ The Satellite Snoopers

The USA, Japan and several West European nations all want to be the first to succeed commercially in selling 'snapshots' of the world from outer space. These snapshots are bundles of data resulting from sensors on satellites picking up radiation reflected from the Earth. At their most straightforward, the data equate to detailed photographs of areas of the Earth. At a more complex level, the sensors tune into radiation other than visible light which, suitably translated, gives useful knowledge about conditions on the surface of the globe or in the atmosphere.

The satellites revolve around the Earth at a height of a few hundred kilometres gathering information which may be put to many uses. From the radiation picked up by the spacecraft, scientists and engineers deduce the sites of minerals deposits on land or the positions of fish breeding areas at sea; they follow the growth in crops and the routes taken by oil slicks and icebergs; they monitor the weather; in industrialized countries, the Earth watchers calculate heat losses from factories; in the Third World, they spot how soil erosion is despoiling agricultural land; the scientists map rivers to work out the most appropriate places for dams and, by spotting geological fault zones, predict the location of earthquakes.

The story has a less peaceable side. With the information gathered by similar remote-sensing satellites, military chiefs can pinpoint the positions of enemy troops, or can work out, to within a few metres, the sites of installations such as airfields that are the targets for nuclear missiles. Observation satellites play a big role in the panoply of modern warfare, but we shall leave the details of the military uses of outer space until Chapter 6.

What is this radiation that the satellites are so successful in picking up and which humans on Earth are so adept at turning to their advantage? In space, nearly everything is black. The darkness is punctuated only by pinpricks of light either emitted or reflected from stars or planetary bodies, but this is to consider only that part of the electromagnetic spectrum to which our eyes are sensitive. This 'visible radiation' has wavelengths in the region 0.4 micrometres to 0.8 micrometres (one micrometre is one millionth of a metre). Many other kinds of electromagnetic wave have wavelengths outside this region.

Included in the list (Table 9) are infrared waves, the sort emitted by furnaces and electric fires, which give rise to heat. Others are X-rays and the relatively long-wavelength (short-frequency) radio waves with which messages are sent around the Earth, including those sent via communications satellites.

Although our eyes see only a small fraction of them, electromagnetic waves from the breadth of the spectrum traverse the universe. If engineers were ever to plot the paths of these waves, they would make a complex series of tracks, like those from grapeshot fired by a multitude of scatter guns positioned in every quarter of the heavens.

Most of this radiation does not concern us. Men and women are parochial creatures. They are interested mainly in the electromagnetic radiation emitted from, or which pertains to, the Earth. A satellite high in the sky is a superb vantage point from which to collect this radiation, not just light in the visible band but other

Table 9. The electromagnetic spectrum

Wavelength				
0.0003 nm	Gamma rays	invisible	increasing wavelength ↓	increasing frequency ↑
0.3 nm (nanometre)	X-rays			
	Cosmic rays			
30 nm	Ultraviolet			
0.3 μm (micrometre)	Blue	visible		
	Green			
0.8 μm	Red			
3 μm	Infrared	invisible		
0.3 cm	Microwaves (satellite communication)			
30 cm	Radio waves for TV transmission			
30 m	Radio waves for speech			

waves too. Also, satellites may rotate about the Earth so they cover different parts of the globe. A satellite system needs an efficient way in which its radiation data is transferred to the Earth, where it can be mixed with information collected from other sources and interpreted. Engineers normally code the data from the spacecraft in radio waves which are gathered by receiving stations over which the satellite passes.

These remote-sensing satellites collect visible light, in effect, by taking photographs of the Earth below. The 'cameras' on the space vehicles do not normally contain film. Instead, light reflected from objects – trees, crops, water, buildings and so on – is turned into an image by an electromechanical device. The images are stored as groups of digits in an electronic memory before being sent to Earth. Usually, the satellite picks up radiation from outside the visible band as well. It is useful, for example, to collect infrared waves. This was first realized during the Second World War when intelligence officers detected camouflaged objects with infrared sensors on aircraft. To us, weapons covered by branches or khaki-and-green cloth blend in with the background, but cameras containing infrared film capture the difference between the radiation reflected by ordinary vegetation and that given off by dead trees or fabric. Hence the resulting images can record hidden installations.

Another well-known use of infrared instruments, for purely terrestrial applications, is to 'see in the dark'. Objects such as animals or buildings continually emit heat in the form of infrared waves. This gives a record of the objects' whereabouts, even when there is insufficient light for them to be seen in the true sense.

Radiation from specific parts of the electromagnetic spectrum is useful in quantifying certain characteristics about the Earth. All substances reflect light or other types of radiation in a unique way, so by measuring the reflectance we may identify the substance. Still better, the radiation from not one but several parts of the spectrum can be collected. Different objects will reflect radiation from each part of the electromagnetic spectrum in a different manner. The full list of these characteristics is known as the 'spectral signature' of the substance. By matching a set of signatures with the data collected from space, a scientist can build up a picture of what is present on the Earth.

For example, a researcher might want to discover the proportion of healthy wheat in a particular country. The obvious way to gather the information (assuming that no government statistics are available) is to telephone all the farmers and ask them to assess the proportion of their crops that is diseased. Even if all the farmers had telephones (which in developing countries they do not) and if they had ready access to the information and were prepared to give it out to a stranger who rings them up, the data would almost certainly be out of date by the time the researcher received it, and even to gather such out-dated information would cost a great deal of time, effort and money.

With satellite remote sensing, however, the picture is different. All the researcher would need would be access to the coded messages obtained from a suitable observation satellite together with the spectral signatures of healthy and diseased wheat (Figure 5). Figure 5 shows that, though some small differences exist in the way the two kinds of wheat reflect light in the visible region, it is in the infrared part of the spectrum where the most marked differences show up. Diseased wheat reflects this kind of radiation far less than the healthy variety. By analysing the remotely sensed data, perhaps with a computer, an accurate description would be obtained of conditions on the ground. What is more, this information could be updated periodically simply by gaining more readings from the spacecraft.

In a similar way, the spectral signatures of rock, water or trees are different from those for both kinds of wheat. The reflectance patterns of wheat are nothing like those for barley. Wheat infested by crop-eating beetles shows a spectral signature at odds with that of the same crop smitten by mildew.

The more detailed the results the researcher requires, the more spectral bands he will want the satellite to take readings in, and the more detailed these readings will need to be. In one celebrated example, with images collected from space of the Imperial Valley in California, scientists identified more than twenty-five separate crops in about 9,000 individual fields. Among the crops were onions, corn, tomatoes, carrots, soya beans, oats and several types of grass.

In practice, the number of spectral bands in which satellites 'look' for radiation varies between four and seventy. In the best-known technique, the satellite contains a device known as a multispectral scanner which contains several electronic detectors (photoelectric

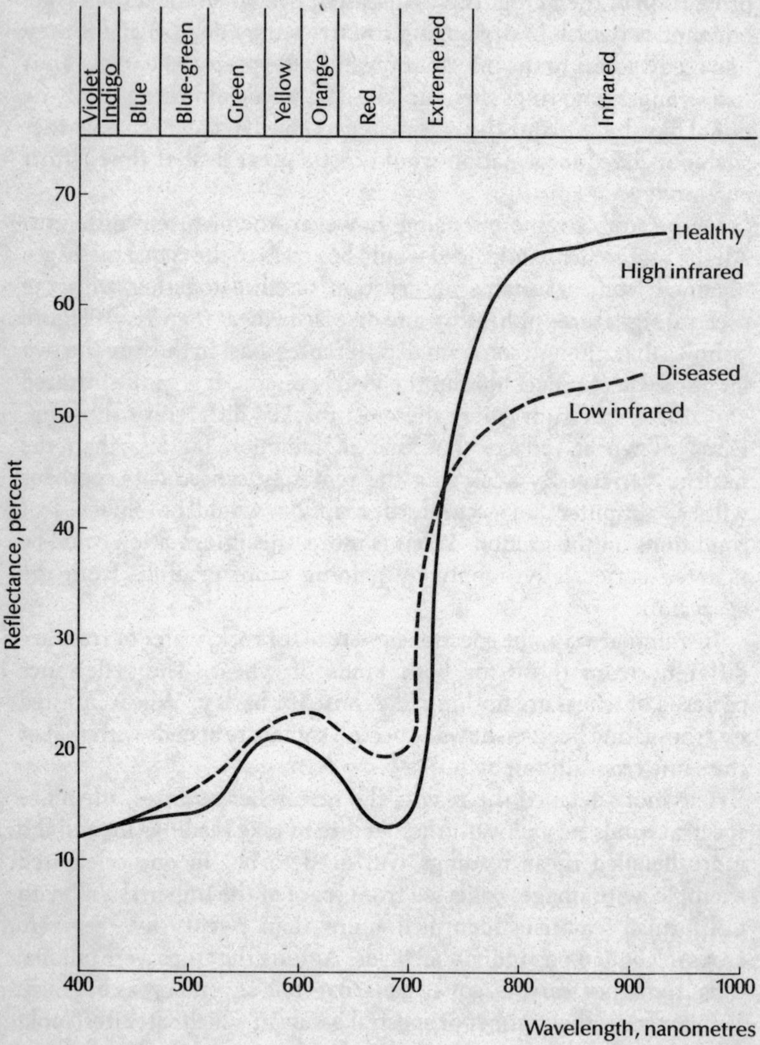

Figure 5. Typical spectral signatures for healthy and diseased vegetation

cells) that respond to radiation in the specific bands in which the satellite operator is interested. An oscillating mirror sweeps across the satellite's line of flight, reflecting radiation from the Earth to filters that split it into the relevant bands.

In this way, electronic images obtained by the detectors are correlated with the precise location on the Earth's surface to which they relate. Once safely on the ground, the satellite information can be translated by computer at the receiving station into a map which plots, say, the whereabouts of maize fields against known physical features on the ground.

Most remote-sensing satellites are in what are called polar orbits. They sweep around the world at right angles to the Equator over the North and South Poles. As the Earth continually revolves about its axis, it follows that the vehicle passes over adjacent swathes of territory each time it completes an orbit. A craft in this orbit about 500 km high will take around 100 minutes to complete one sweep of the Earth, passing over the same spot on the globe's surface every twenty days or so.

The history of remote-sensing satellites is remarkably short and is dominated by events in the USA. However, other countries are catching up fast, much to the concern of American satellite operators. They fear that their lead in satellite imaging could be eroded as other nations strain to turn this activity from a research tool into a commercial undertaking.

After the launch in the 1960s of several experimental TIROS (Television and Infrared Observation Satellites) vehicles, the US government put into space in 1972 the first fully operational observation satellite for civilian use. This was the first of what became known as the Landsat series. Four further Landsat vehicles were launched in 1975, 1978, 1982 and 1984. Of these, only the last two, Landsats 4 and 5, are still operating. The first three satellites in the series were controlled by the National Aeronautics and Space Administration (NASA). Landsats 4 and 5 are, for the time being, the responsibility of another government agency, the National Oceanic and Atmospheric Administration (NOAA). This is an 'operational' organization rather than an agency like NASA that is geared mainly to research. The change reflects the desire of the American government to establish Landsat as a commercial service. In the next couple of years, the government plans to hand

Landsat over to a private company to run as a business operation.

The technology in the first three Landsats remained static. All carried the same variant of multispectral scanner plus another type of sensor called a return-beam vidicon which acts like a TV camera. Thus the resolution (the ability of the spacecraft to discriminate between small objects on the ground) and general performance of the Landsat vehicles failed to improve during the 1970s, just the period when the USA might have been expected to go all out to consolidate its position in remote sensing.

The advance in technology came eventually with Landsat 4, which in addition to the multispectral scanner carries a new instrument called the thematic mapper. This is a sensing device that collects data in seven spectral bands (Table 10) as against four with the multispectral scanner. The instrument has a resolution of some 30 metres, compared with 80 metres with the device on the earlier Landsat vehicles. Thus it can pick out objects on the ground as little as 30 metres in dimensions.

Under the current procedure, Landsat information can be obtained in two ways. The data can be purchased directly from the US government, which runs a Landsat data collection centre in Sioux Falls, South Dakota, where the staff are most helpful in

Table 10. Spectral bands for thematic mapper

Band	*Spectral range* (micrometres)	*Application*
1	0.45–0.52	Coastal water mapping; differentiation of soil from vegetation
2	0.52–0.60	Vegetation measurement
3	0.63–0.69	Chlorophyl absorption band used for discriminating between different types of vegetation
4	0.76–0.90	Determination of biomass content of plants
5	1.55–1.75	Assessment of moisture in vegetation and soil; differentiation of snow from clouds
6	2.08–2.35	Discrimination of rock types
7	10.4–12.5	Soil moisture analysis

sending price lists, catalogues and so on. The cost of data depends on the sophistication required and may not be too expensive – for example tens of dollars for a photographic print of an area of the Earth of 185 km square – whereas a computer tape containing much more detailed information about the same area would cost $730 for an image derived from the multispectral scanner and $4,400 for one obtained from the higher-reduction thematic mapper. (These prices are correct for 1985.) The computer tape would contain a digital representation of the radiation measured by the Landsat craft in several spectral bands, which could then be run through a computer to upgrade it or put it into a different format.

This last processing operation can be far from cheap, though it improves the value of the satellite information enormously. According to one estimate by the US Congress's Office of Technology Assessment, converting a single computer tape to a usable form can cost between $1,000 and $20,000, depending on the sophistication of the processing. This puts customers for Landsat data from Third World countries, for example, at a disadvantage where the computer expertise to 'massage' the data from space in the most appropriate way is not available.

This drawback can to some extent be redressed by the second way of obtaining Landsat information – by setting up a ground station yourself. The receiving equipment that picks up the Landsat signals from space can be combined with the processing hardware that makes sense of the raw radio code. The station plus computer facilities would cost $10 million to $15 million. In addition a 'licence fee' of $600,000 per year is payable to the American government, in return for which it supplies a list of Landsat orbits, plus the codes and the radio channels on which the craft send their information. This makes it physically possible to tune in to the satellite's signals. Not surprisingly, the people who set up such arrangements are governments of relatively large countries who have this kind of money to spend. Outside the USA around a dozen receiving stations are entitled to 'listen in' to Landsat transmissions. They include listening posts in Canada, Australia, Sweden, Italy, Brazil, Argentina, Thailand, South Africa, India and Japan. Saudi Arabia, China and Pakistan are also due to set up stations in the late 1980s.

As a result of these arrangements, research scientists the world over have had a wonderful opportunity to use Landsat data in many different disciplines. It is difficult to think of another scientific technique that has proved itself so versatile. In Central America, workers from the US Geological Survey used Landsat data to predict the eruption of the Fuego volcano in Guatemala in 1973. They did this by correlating knowledge about geophysics with Landsat information which showed up, for example, fault structures in the Earth surface. In the USA itself, Landsat images have helped in studies ranging from the assessment of spoiled mining land in Missouri to the inventory of forests in Idaho. Economic analysts regularly harness crop information from the satellites to evaluate the likely harvests of countries such as the Soviet Union where accurate forecasts are not normally available.

Scientists have produced soil maps of places such as northern India to determine the sites where particular kinds of vegetables and other crops can be expected to grow best. Satellite data produced the first detailed maps of Bolivia. With this technique, researchers have plotted where to build new roads in Ethiopia, monitored soil erosion in Lesotho and the Sudan and explored the terrain for tin deposits in Thailand and Indonesia. In one project sponsored by the United Nations' Food and Agriculture Organization, Landsat information provides an early warning of swarms of crop-destroying locusts in parts of Asia and Africa.

Similar projects have forecast floods in countries as far apart as Poland and the Philippines and mapped hydrological schemes in Burkina Faso and Uganda. In the UK, workers at the Institute of Geological Sciences are trying, with the help of satellite data, to earmark deposits of metals even when they are deeper than most mines at more than 3 km under the ground. Other surveys from space have been helpful in minerals projects all the way from Egypt to South Australia.

The list seems impressive, yet not everyone is happy. For one thing, countries in the Third World, which are frequently the beneficiaries of remote-sensing studies, are often aggrieved that the researchers who do the work are from the relatively well-off countries. Insufficient training in space science is available in the Third World to ensure that enough people from these countries really understand remote-sensing technology. When a specific study is finished, after six months or a year, the researcher from the industrialized nation goes home to start on something else, leaving no one from the developing nation to carry on with the work. As a result, Third World countries may find themselves the subject of a stream of one-off development projects co-ordinated by international agencies, while continuing to miss out on the long-term benefits of remote-sensing technology.

In the USA, too, there is widespread concern about Landsat, though for very different reasons from those of the developing nations. First, the USA loses a lot of money in its dealings with other countries over Landsat. In a bid to reduce the deficit, in 1982 the government raised its prices for Landsat data more than twofold. Before this, it was obtaining only about $6 million annually from foreign buyers of the information.* This has to be set against the tens of millions of dollars that the Landsat service costs the US taxpayer each year. People in American government circles also feel that enterprises such as Landsat should be run not by state agencies but by private concerns. This was believed by some elements in President Carter's administration and still more so by the more right-wing government of President Reagan. According to this argument, a private firm would show more imagination in operating the remote-sensing satellites, and might even make a profit.

Difficulties may lie ahead, however, as the US Congress's Office of Technology Assessment explained in a report on American space policy in June 1982:

> Continuing to develop land remote sensing . . . represents a certain economic risk for the government or the private investor. If the market cannot sustain the investment, the losses could be great. However, the government and private industry have already committed more than $1 billion [over some ten years] to the Landsat venture. To fail to make the best use of these sunk costs represents a considerable loss as well.

A way out of the impasse appeared to come in early 1983. President Reagan gave his long-awaited formal approval to a proposal to sell off Landsat to private industry. Under this directive, the Department of Commerce was to start serious discussions with firms to

* In the twelve months to October 1983, the US government collected $14 million in Landsat sales, offsetting operating costs of $21 million. Estimates by the Department of Commerce for 1984–5 point to sales staying roughly constant with costs rising to $35 million. This does not include the $200 million–$300 million that is the cost of a new Landsat vehicle, required every three or four years as the older satellites wear out.

find a buyer. But controversy reared on account of Reagan's notion that, to make the Landsat deal attractive enough to industrialists, the US government's meteorological satellites would become part of the sale as well.

Such weather vehicles operate rather similarly to the remote-sensing craft. There are two basic kinds of meteorological satellites. The first travel around the Earth in a polar orbit, covering the whole of the globe. They pick up, with infrared sensors for example, information about the amount of heat in the atmosphere. Alternatively, weather craft may hover over one point on the Earth's surface, in a geostationary orbit 36,000 km above the Equator. In this case, the vehicles relay to ground stations far less detailed information about a very large area of the globe. From this can be extrapolated broad-brush general knowledge about weather trends. Data from both types of satellite can be combined with other kinds of weather information, from ground-based monitoring centres for example, to provide detailed knowledge about meteorological conditions. In the USA at present, NOAA runs the civilian weather satellites in addition to the Landsat craft.

Initially, the chief industrial participant in the discussions over satellite 'privatization' was Comsat, the Washington firm that already runs several communications satellites and has an eye on establishing a direct-broadcasting TV service from outer space (see Chapter 4). Paul Maughan is the head of Comsat's Earthstar team, the group that wanted to take over the government's satellites. Earthstar first suggested that a 'job lot' of both the Landsat and weather systems would be an attractive proposition for the private industrialist. Maughan said that Comsat, if it were allowed to take over the systems, would reduce staff and put more modern technology in both new satellites and ground stations. With data from the meteorological satellites, the firm would set up commercial weather services, selling detailed forecasts to farmers for example.

Comsat offered to pay the government $300 million for the five weather and one Landsat satellites already in orbit, 'spares' and other space hardware on the ground plus Earth stations. Comsat also proposed that the government should guarantee to buy data from the new operation for fifteen years, agreeing in advance to pay out some $325 million per year. This would continue the arrangement under which federal and state agencies are currently big users of Landsat for weather information. However, the proposal was seen in some quarters as rather cheeky, to put it mildly. It could entail the government in handing over more cash for the data under the new arrangement than it does at present through ownership of the satellites via NOAA.

This was not the only reason for the proposed satellite sale attracting criticism. According to some views expressed in Congress, the US government was planning to hand over control of a vital public asset to businessmen bent only on making a fast buck. To sell off the weather satellites could also jeopardize the arrangement under which the USA 'swaps' data from its meteorological craft, via a telecommunications link, with the other nations that operate similar satellites. These countries are Japan, the USSR and the eleven countries of the European Space Agency. The interchange of information in this way is co-ordinated by the United Nations' World Meteorological Organization based in Geneva, and ensures that the weather can be monitored on a global scale to provide a better understanding of weather patterns.

Towards the end of 1983, the critics of the sale of the weather vehicles won the day. An important vote in Congress confirmed that politicians would stop the President going ahead with the scheme; but the sale of Landsat was very much on. In June 1984, the Department of Commerce announced that it had narrowed down the list of private-sector contenders for Landsat to two consortia: EOSAT, which is led by Hughes and RCA, and a group in which Eastman Kodak is the senior partner. After further discussions, Eastman Kodak dropped out, leaving EOSAT as the only contender. At the start of 1985, EOSAT was still without a formal contract to take over Landsat, however. This was because of lack of agreement within the government on the level of subsidy from federal funds to which EOSAT would be entitled.

The two other companies in the EOSAT venture are Computer Science Corporation (which would handle ground processing of data) and Earth Satellite Corporation, which would look after marketing – a factor totally neglected in Landsat operations to date. Dr Charles Sheffield, Vice-President of Earth Satellite Corporation, says that, with proper attention to setting up a marketing operation, sales of Landsat data could come to $100 million a year by the mid-1990s. If his group won final approval from the Department of Commerce, it would spend several million dollars on new satellites to follow Landsat 5 and ground support services, including new Earth stations. The first new satellite built by EOSAT, Landsat 6, would enter orbit in 1988, with Landsat 7 following in 1991. There would be an accent on new technical features aboard the satellites, in stark contrast to the somewhat leisurely technological development of Landsat during the 1970s. For instance, if it gets the go-ahead from the government, EOSAT will introduce sensors that take measurements in no fewer than thirty-two spectral bands in the visible and infrared range of the spectrum. Customers who wanted the satellite to take measurements in a specific band, for instance to find out details about a particular kind of vegetation, would instruct the operator, which would program the space vehicle in advance as to the details of the way it 'looks' at the ground.

An important issue will be the cash support that whichever group wins the government approval obtains from the US taxpayer. It looks almost certain that Congress will have to guarantee a proportion of the development funds for Landsat – even when the system is under private control – as a way of continuing a strong US presence in remote sensing. The level of this subsidy could run to about $50 million a year for the first five years after a private concern takes over the system.

Not everyone agrees with this strategy. Another proposal is that the USA should shed no tears over the demise of Landsat. It should instead let the system, long in the tooth and increasingly moribund, die a natural death. Private companies should then get on with the job of creating a new remote-sensing system with brand new technology.

This is the attitude taken by a company called Space America, based near Washington DC. The company has unveiled plans to build its own remote-sensing satellite. This would cost only $30 million, a tenth of the price of a new Landsat vehicle, according to Diana Josephson, a graduate in law from Oxford University who is Space America's President. The spacecraft would use modern technology – sophisticated electronic sensors with no moving parts – instead of the multi-spectral scanners carried by Landsat. Ball Aerospace has agreed provisionally to build the spacecraft, with Honeywell lined up to provide the sensor.

If Space America's plans go ahead, it will probably launch its vehicle into an altitude some 900 km above the Earth. At the beginning of 1985, it was talking to foreign companies who would put up the cash for the satellite. Space America could then operate it on their behalf. The company may decide to use a rocket owned by Space Services, a group of Texan entrepreneurs who want to challenge the dominance of government agencies in the launch business. Space Services plans a rocket called Conestoga, built mainly from parts of existing launchers, which would take satellites into orbit for less than the price charged by NASA.

In remote sensing, one of the principal spurs for people in the USA to *do* something – even if they are not quite sure what – is that other countries are proceeding quickly with their own plans for satellite imaging. If the USA fails to pull together its operations in this area of technology, it may find that other countries win the race to commercial remote sensing.

The biggest threat is from France, which from 1986 plans to offer a remote-sensing service along commercial lines. Spot Image, a company owned by the French government and private shareholders, will operate the service. Initially the company will have one satellite, with a second to follow in 1987. The total investment will be around $400 million, most of it government money. CNES, France's national space agency and the equivalent to NASA, owns one-third of the shares in Spot Image, the rest being held by private or French-government-owned firms. The Spot vehicles will have better resolutions than their Landsat counterparts. They will 'see' objects as little as 10 metres in dimension. What is more, the solid-state sensors aboard Spot will be steerable and will provide stereoscopic images of the ground. Hence the French will give customers information about land not just in two but in three dimensions. This could be particularly useful in mountainous areas.

Governments or companies wanting the data will normally operate their own ground stations. Spot Image's representatives are trying to negotiate contracts with nations who already own Landsat receiving hardware, which, with a modicum of electronic twiddling, will be able to tune in to Spot as well. Unlike the arrangement with Landsat, however, these organizations will not have to pay a flat annual 'licence fee' to the French. Instead they will telex to Spot Image's headquarters in Toulouse instructions about the kind of information that they want the spacecraft to pick up. For instance, a country in Asia might want the vehicle to 'look' at a particular area of paddy fields as part of an assessment of the level of the next rice harvest. The information would be channelled from Toulouse to a computer on board the satellite. The computer would 'remember' the instructions, telling the sensors to switch on when the vehicle is over the paddy fields, and in which direction to point to photograph the views required.

The French firm's prices will be roughly equivalent to those for Landsat. In 1984, it announced it would sell a reel of tape containing data about an area of Earth measuring some 60 km square for about $1,000. By the late 1980s, Spot Image estimates it will be selling data worth around $50 million per year around the world. A lot of this commercial activity may be at the expense of American remote-sensing organizations which, by then, may be left behind in the race.

Paradoxically perhaps, the fact that the French satellites will have a high resolution is causing the company problems. Some countries, particularly in the Third World, fear that satellites that can take such detailed 'snapshots' represent the space-age equivalent of the nosy parker. These countries are worried about the erosion of privacy resulting from satellites that collect data from outer space. According to this argument, companies or governments could gain access to other countries' secrets by buying data from firms like Spot Image. The secrets are not pieces of paper, but represent information about a country's land area or other resources. Thus a minerals company could obtain satellite information from which it may learn that there is a fair chance of obtaining a specific metal in a particular part of the world. This may give the company an insight denied to the people actually living in the area, who may be literally sitting on a goldmine. The firm could then buy the land at low prices, to the disadvantage of the country concerned, which would be ignorant of the reasons for the deal.

Possibilities such as this do not seem too remote to Third World nations, who may already be suspicious of the activities of the richer countries. The Soviet Union, too, says it is worried about satellite snooping. It has proposed that data with a resolution of less than 50 metres should be subject to some kind of international safeguards. For instance, countries to which the satellite information refers would have the right to veto sales to outside parties. Most Western countries, however, do not want clamps on the sales of satellite data. This, they fear, could hold up the development of commercial opportunities in this area. The arguments of the Third World and the USSR have not impressed the French firm. It intends to follow the lines of the US government in its attitude towards trade in data. Spot Image will sell Spot information to anyone in the world who asks for the data, with no questions asked.

The forward-looking plans of Spot Image are not the only reasons for the unease of Americans interested in setting up a remote-sensing industry. Several other countries plan satellites which will use a new kind of technology to provide photographs from space that may turn out more useful than the traditional kind. These vehicles contain radar devices which direct high-frequency radio waves at the ground and pick up their reflections. Such satellites are called 'active' because, unlike Landsat and Spot, they do not rely on picking up natural radiation in a 'passive' manner.

Radar satellites may 'see' objects on the surface of the globe that conventional craft have difficulty in identifying. They can also penetrate clouds – radar is oblivious to the water vapour in the atmosphere which absorbs ordinary light and infrared waves.

Japan will probably be the first to enter the 'radar satellite' field. It plans to launch, in 1986, a vehicle called MOS-1 which will scan the oceans. The satellite will carry three sensors and a data collection system. Two of the sensors will collect radiation reflected from Earth in the visible and infrared parts of the spectrum, while the final device, known as a microwave scanning radiometer, will direct radar waves at the Earth and collect them on the rebound. The craft will obtain data about the heights of waves, weather conditions and so on which will assist the country's fishing industry, the biggest in the world. This could be the first of several remote-sensing satellites that Japan will launch over next few years, and which could offer a commercial service by the end of the decade.

The eleven nations of the European Space Agency are combining to put into orbit a similar satellite in 1988. The main purpose of this craft, called ERS-1, will again be to look at the oceans. It will gather data about wave heights and speeds and also measure water vapour in the atmosphere. The craft could be used commercially, sending its data back to Earth stations which would then prepare detailed weather forecasts and assessments. These would go, for a fee, to organizations such as shipping firms or oil-platform operators. Scientists could also interpret signals from the satellite to indicate the positions of oil slicks or fish breeding areas.

The final contender in the area of radar satellites is Canada, which wants to launch such a vehicle to watch out for icebergs and other ice movements in the country's arctic regions. Canada's spacecraft, called RADARSAT, would also provide data for general oceanographic studies, for example information about wave heights, and also identify crops and geological features over land areas.

All these satellites are of the same lineage which started with a vehicle called Seasat, launched by the USA in 1978. Seasat was the first civilian radar craft. Unfortunately for all concerned, it lasted for just a few weeks before a fault prevented the craft from returning any more data to Earth. Seasat's life may have been short but it was fruitful. The data that the vehicle sent back have proved extremely useful to researchers in the area of radar remote sensing. Radar images are very difficult to interpret; on such an image, all you can see, often, is a mass of lines. To work out their meaning is like learning a foreign language, say satellite experts. The data have to be matched with what is known about the Earth – this is called establishing 'ground truth'. Scientists continue to pore over images from Seasat, even though the satellite itself ceased operation long ago. If the new radar satellites succeed, researchers will have to be truly thankful for the experience gained by the ill-fated Seasat.

The USA plans to follow Seasat with two more radar satellites, to be launched either in the late 1980s or early 1990s. The Navy Remote Ocean Sensing System (NROSS) should enter orbit in 1989, a joint project between the US Navy, NASA and NOAA. Data from the vehicle will be sent initially from the satellite, as it orbits the Earth, to a military base in Monterey, California. There information will be extracted which will be useful for defence applications, for example on the movement of Soviet warships, before sending data to civilian stations. The NROSS will carry four microwave instruments to provide what its proponents hope will be a capability to keep a close watch on the oceans in all weathers.

The second satellite, called TOPEX, is a NASA vehicle which should provide information about the height of waves to within 2 cm, an extraordinary accuracy. This should enable scientists to gain further understanding of the characteristics of ocean currents, leading to a greater knowledge about, for instance, the effects of wind and waves on fish breeding patterns and the dispersion of pollutants.

Also in the planning stage in the USA is a unified global venture to take readings from space in order to understand the Earth better. The scheme would co-ordinate a number of programmes involving different satellites and a range of countries. The satellites would include the Earth-sensing vehicles discussed earlier in the chapter, whether operated by commercial organizations or governments, together with meteorological satellites and other craft which take readings of the atmosphere. The programme would integrate scientific exercises covering several different disciplines – Sun observation, oceanography, studies of the atmosphere, climatology, plate tectonics and mineralogy – to build up a 'holistic' view of what is happening to the Earth.

Both the USA's National Academy of Sciences and NASA have started discussions on such a venture. The talking is likely to go on for some time yet and on an international scale. The National Academy of Sciences has labelled its scheme the International Geosphere–Biosphere Programme, while NASA prefers to talk of a venture in Global Habitability. If other countries agree – and there is a good chance they will, bearing in mind the level of international co-operation already in remote sensing generally – then the programme is likely to start up in the late 1980s. John McElroy, a top official in NOAA, summed up in a paper in 1984 how a project along these lines could provide valuable planning information for world governments:

> The advances in remote-sensing technology provide a powerful approach to the strengthening of the scientific basis for the long-range prescriptive forecasts of events or trends, either desired or undesired, and allow the evaluation of alternative strategies to reach the desired outcome . . . There is no doubt that remote sensing is one of the most exciting and controversial fields in the entire space programme. It is also true that the assessment of the total Earth environment is one of the most important and challenging tasks of that programme also, and that it is not a task for a single nation or even a few nations, but one requiring and deserving the best talents of the entire globe.

Important advances in remote sensing are likely as a result not just of flights of unmanned satellites but of vehicles that take people into space. Conventional satellites are generally fairly small and do not have room for the largest, most powerful instruments. Manned craft generally have far more space for sophisticated hardware that can take precise views of the globe or accurate readings of the atmosphere. Take, for instance, the first flight in November 1983 of Spacelab, a re-usable laboratory built by the European Space Agency which fits inside the cargo bay of the shuttle. Spacelab has many applications in areas of materials processing and low-gravity experiments (this will be discussed further in Chapter 6). It also acts as a carrier for remote-sensing instruments. On the first flight, it took into space hardware that made the first detailed measurements over the whole of the globe of trace elements found in the atmosphere between 40 km and 100 km above the Earth. Before the flight, scientists had only a poor understanding of this part of the atmosphere. Researchers obtained readings of gases such as ozone, methane, carbon dioxide and carbon monoxide. By repeating the experiments on later Spacelab flights, scientists may be able to keep track of gaseous pollutants and build up models for the ways in which the materials disperse naturally through the atmosphere. Workers may also find they can monitor in detail variations in the ozone layer. Any changes in this layer may affect life on Earth, as the gas acts as a barrier to harmful radiation from the Sun.

NASA planners hope to return the experiments to space once every one or two years on Spacelab flights as part of what they call the Earth Observation Mission. The first such flight is due in August 1986. Not only will Spacelab offer an opportunity for data acquisition with particularly powerful instruments, but the fact that people are on board provides a remote-sensing mission with greater flexibility. For instance, astronauts on the first Spacelab voyage were able to point a big, West German-built instrument called the Metric Camera at regions of the Earth that they found particularly interesting. Another bonus is that data-gathering devices on manned vehicles can be returned to Earth for recalibration prior to a new flight, an important factor when the readings need to be especially detailed. In contrast, the hardware on conventional unmanned satellites is, of course, never returned to Earth.

As flights continue, the hardware taken above the atmosphere will become more sophisticated, providing unusual results. Thus the second mission of the space shuttle Columbia in November 1981 carried a radar instrument that inspected the Earth from a height of 300 km. The results from this were staggering. Researchers believe that the radar images of the Sahara desert exhibited properties similar to those obtained with X-rays; the images showed what was present several metres beneath the surface. They displayed physical features such as river terraces and desert basins buried by sand. Subsequent investigations showed that these interpretations were correct.

This technology could be of huge importance to archaeologists. It is possible that radar images from space may give them an understanding about artefacts of previous eras. It may be that, with better resolution and interpretation of data, researchers can discover small objects hidden underneath the ground. It would be interesting indeed if one of the most modern technologies enabled mankind to stop for a moment and gain more of an insight into what was happening centuries if not millennia ago.

6 ☆ Space War

Control of space will be decided in the next decade. If the Soviets control space they can control Earth, as in past centuries the nation that controlled the seas dominated the continents.

President John Kennedy, 1960

There would not yet have been a single astronaut or cosmonaut in orbit without the help of rockets initially developed for military purposes.

Wernher von Braun, 1965

We have spent money, dollar after dollar and billions and billions, for weapons whose only consequence is to kill people. Now we have within our capability the possibility of developing weapons whose only real role in the world is to kill the things that kill people.

Senator Malcolm Wallop, 1979

Like approaching storm clouds, [it] threatens to cut off all rays of hope: this is the increasing and rapidly escalating militarisation of outer space . . . An arms race in space would increase the areas and the potential for confrontation, adding a new dimension to the human destruction that would stem from it.

Javier Perez de Cuellar, Secretary General of the United Nations, 1982

These statements sum up the state of affairs in what is, with little doubt, the area of space activities in which most money is being spent and in which technological advances are taking place at the greatest speed. The exploitation of space for military reasons is a fact of life for the world's biggest powers.

President Kennedy's remarks illustrate the basis for this. Over the centuries, Earth-bound armies have sought to hold mountainous areas because they are good defensive positions or useful points from which to mount an attack. In a similar way, space acts as the ultimate in 'high ground' as a place from which force can be wielded. Space technology is bound to figure large in any strategy by a world power for safeguarding its Earth-based interests.

The comment by Wernher von Braun, doyen of the rocket

pioneers, recognizes the driving force of military imperatives in the general development of space technology. As was clear from Chapter 2, military chiefs have always been interested in rocketry, from the days of the Chinese 'flying arrows' to the forays with gunpowder-driven munitions in the early 1800s by people like Sir William Congreve.

This development reached its crescendo during the Second World War, when engineers working for the German Army built rockets that, for the first time, reached the upper limits of the atmosphere, almost into space. The V2 weapons, designed at a missile station in Peenemünde, terrorized the citizens of London in the closing stages of the conflict. Not only this, they laid the base for the much more powerful instruments of war that were to emerge after the 1939–45 conflict ended.

Both the USA and the Soviet Union quickly latched onto the concept that a ballistic missile that cruises through space on its way to a target could be a fearsome armament indeed. From the bombed-out laboratories and missile sites of war-torn Germany, the superpowers recruited both ideas and people. They converted the V2 weapons built at Peenemünde into missiles with far greater ranges, of up to several thousand kilometres. The final touch was to add as a warhead not simply the high explosive that the V2 missiles had carried but something far more deadly, a nuclear bomb.

In this way was born the intercontinental nuclear missile, a device which for forty years has threatened a large proportion of the industrialized world with obliteration and, as a result of its massive power to destruct, set in motion between the forces of the NATO Alliance and the Warsaw Pact the successive periods of cold war, *détente* and back to cold war.

But nothing stands still in war, nor in theories for fighting it. Engineers from laboratories and weapons establishments from around the world – and we must remember that military work accounts for a quarter of the Earth's research and development budget – have produced layer upon layer of ever more ingenious hardware to support and make more effective the missiles themselves. Much of this hardware exploits the useful qualities of space as a place relatively difficult to get at and which provides a superb vantage point from which to wage, or threaten, hostilities.

The first layer of hardware consists of satellites of various forms.

Three-quarters of the 3,000 or so vehicles that the world has injected into space since 1957 have had military applications, mainly in support roles for missiles (Figure 6). They monitor enemy installations, provide an early warning against missile attacks, supply communications links and can even add to the accuracy of missiles, by furnishing them, while they are in flight, with information about variations in the Earth's gravity.

In recent years, the USA and USSR have tried to develop still more sophisticated space-based equipment for backing up their military efforts. This comprises the antisatellite or antimissile weapons that may soon be based in orbit around the Earth. The hardware could consist of particle-beam or laser guns which shoot down enemy installations, in the style of Buck Rogers or characters from *Star Wars*. Both the superpowers are striving to deploy this weaponry by the 1990s. The weapons can be regarded in three ways.

First, the hardware could function as an undeniably hostile system that would give the nation that perfects it a supreme advantage in any world conflict. The beam guns could fire the first salvos in a war, winning the contest at a very early stage by destroying the satellites that form the opposition's military infrastructure.

Second, the weapons could remain hostile systems but never be used. They would become the ultimate deterrent, warding off the threat of hostilities by the knowledge of the massive retaliation that would follow any aggressive move by the other side.

Third, the weapons could be thought of as purely defensive systems. This is where Senator Wallop's statement, quoted at the beginning of the chapter, becomes relevant. A series of laser guns deployed around major cities could render these places safe from attack by enemy missiles. This assumes, of course, that scientists can solve the horrendous physical problems of first constructing the hardware, and then ensuring it can be accurately fired at the enemy's intercontinental missiles before they reach their targets.

In March 1983 President Reagan announced that the USA would start to develop such a defensive system. As a first step, the Department of Defense is due to spend $26,000 until 1989 on research into technological elements required for the system. On the one hand, if it is ever developed, such *Star Wars* technology will make the citizens of major American cities sleep a little easier in

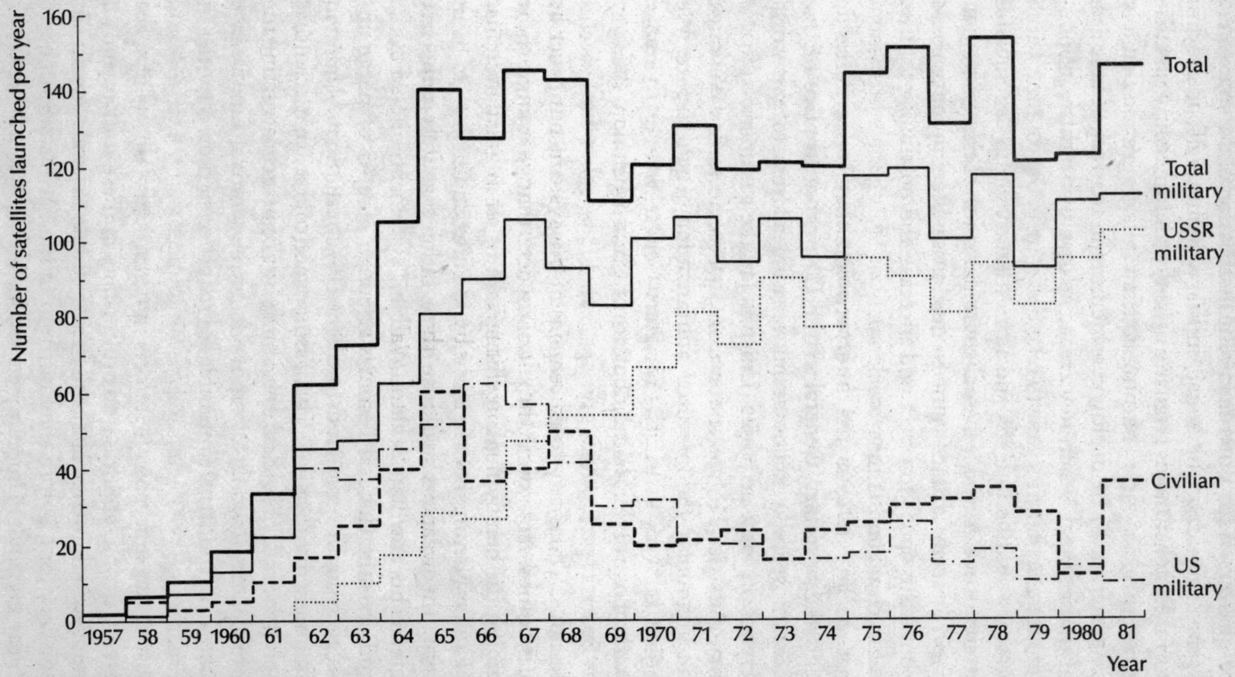

Figure 6. Military and civilian satellites launched between 1957 and 1981.
Source: B. Jasani, ed., *Outer Space – A New Dimension of the Arms Race*.

their beds; but on the other hand, the installation of the weaponry, for all the rhetoric about it being 'defensive', could well add another twist to the ratchet that is cranking up the arms race. An antimissile system of this kind, or, to be more accurate, the realization that just such a system may be deployed some time in the future, acts as a signal to the Soviet Union to increase its own war-fighting efforts, on the grounds that, if it fails to take this course, the country's arch adversary would be one step ahead in military terms, an advance that could prove crucial if ever the world is on the brink of war. In short, what seems at face value to be a peace-keeping move could well have the opposite effect and increase the possibility that an all-out nuclear conflict may break out.

That, at any rate, is the interpretation that Javier Perez de Cuellar, the Secretary General of the UN, chose to place on the build-up of space weapons at his opening address to the world conference on space activities (Unispace) in the summer of 1982 in Vienna, Austria. He spoke for many when he said: 'We must oppose vigorously the increased militarization of outer space. We have time, but very little.' The Secretary General went on to invoke the spirit of one of the landmark pieces of world legislation, the 1967 Outer Space Treaty, which says space should be solely for peaceful applications. The treaty bans weapons of mass destruction, such as nuclear bombs, from outside the atmosphere though it stops short of vetoing the deployment in this sphere of all military systems. 'The United Nations has the duty,' said Perez de Cuellar, 'to be in the forefront of those who choose peace, rather than war, in this last and greatest of frontiers of the Universe.'

The Secretary General's words, although greeted by polite applause, were hardly enough to cause the USA or the Soviet Union to rethink their military policies for outer space. The two big powers have committed a great deal to building up their force levels outside the atmosphere. It would take more than a few words at a conference to make these countries contemplate a reduction of their activities in this area.

Most of the superpowers' efforts in the military aspects of space consist of putting satellites into orbit. Many of these craft are only a few hundred kilometres above the Earth (Table 11) and they do not last for very long before they burn up in the Earth's atmosphere. Each year the Soviet Union injects around eighty-five military craft

Table 11. Different types of military satellite

Satellite	*Altitude (km)*
Photographic reconnaissance	
Big Bird	180–290
KH-11	240–530
USSR	180–350
Electronic reconnaissance	
US	480
USSR	500 650
Ocean surveillance	
US	1,100
USSR	250
Early warning	
US	36,000
USSR	688 × 39,000
Nuclear explosion detection (US)	110,000
Meteorological	
US	36,000
USSR	900 610
Communications	
US-SDS	250 × 39,000
US-FLTSATCOM	36,000
USSR-Molniya	440 × 40,000
USSR-Cosmos	1,400
NAVIGATION	
US	20,000
USSR	1,000

Source: B. Jasani, ed., *Outer Space – A New Dimension to the Arms Race.*

into space and the United States fifteen. The lower number for the USA is because its vehicles normally see service longer than those of the USSR. Soviet craft may last for less than a year before they are replaced. Military satellites have a peculiarly ambiguous role. Most of them carry no weapons, which would appear to make them non-aggressive. In fact, the only way that a typical military spacecraft could injure someone or something is by falling out of the sky and delivering a mighty thump to whatever it happens to land on.

A related theme is that some of the spacecraft may actually help to preserve peace. These are the surveillance vehicles that watch out for missiles that either the USA or the USSR deploys. If either side steps out of line, for example by installing more weapons than are permitted under arms-control agreements, then the opposing country knows about it almost immediately, thanks to its continually orbiting network of 'spies' in the heavens.

There is a counter-argument, however, according to which the seemingly benign military satellites and the obviously murderous missiles of destruction are inextricably linked. The weapons could not function without the massive command-and-control and information-gathering network that the superpowers' array of satellites makes possible. Bhupendra Jasani, of the Stockholm International Peace Research Institute in Sweden, sums up: 'I cannot for the life of me justify that, for instance, a navigation satellite that can guide a weapon to within 10 metres of its target represents a peaceful activity.'

The most common type of military satellite is the surveillance craft. These are defence chiefs' 'eyes' and 'ears' in space. The vehicles have a variety of sensors that obtain information from the ground, with TV cameras for instance; 'listen' to radio transmissions with electronic mechanisms; or watch out for events in the atmosphere or in space itself, for example the aftermath of nuclear explosions. The USA launches about three photographic reconnaissance craft every year; the USSR's tally is thirty-five or so. Again the discrepancy is because the USA's vehicles last a lot longer, two or three years as opposed to forty days.

Information from photographic satellites is useful in pinpointing enemy troop movements or plotting the sites of sensitive installations. The vehicles can also provide direction-finding information for missiles. For example, the USA's cruise missiles in Western Europe will be guided to their targets by digital maps stored in the missiles' own electronic memories; the maps will be updated periodically with data from spy satellites.

In the days of the cold war, the United States in particular had ample reasons to develop spy satellites. To find out about the USA's ballistic missiles, the Soviet Union had only to buy government maps freely available in Washington. For the USA it was not so easy. No Westerners are allowed anywhere near Soviet missile bases and official maps are often falsified to show weapons silos many kilometres from their true positions.

The photographic surveillance craft launched by military agencies resemble the civilian mapping vehicles discussed in Chapter 5. Both kinds of activity can, of course, trace their origins back to the reconnaissance operations carried out by aircraft during the Second World War. Sorties by aeroplanes, which carried either cameras or simply observers who scanned the ground with the naked eye, established enemy targets and were often preludes to bombing raids. Interestingly, it was just such a mission by the RAF, in June 1943, that established for British intelligence forces the nature of the engineering work at Peenemünde in building the V2, the descendants of which were to make the huge expansion of military surveillance operations possible in the decades to come.

The difference between military and civilian surveillance satellites is the greater accuracy of the former. The instruments on military vehicles are more sensitive and, secondly, the spacecraft fly nearer to the Earth, so they can take a closer look at anything interesting. Some of the big powers' spacecraft can differentiate between adjacent tiles on a roof-top or, so it is claimed, read a car's number plate. The resolutions needed to identify particular military installations are given in Table 12.

The spy craft used by the military return their information to the Earth in two ways. They can record pictures on a film which is then jettisoned attached to a parachute. The film capsule is later picked up from the sea or, in a more sophisticated approach, it is snatched by an aircraft while still in motion through the air. Satellites that return film to Earth in this fashion are popular with the Soviet Union. The USSR may launch specific spy craft, which are about the same size as the Soyuz modules that ferry cosmonauts to and

Table 12. Resolution (in metres) required for military interpretation tasks

Target	*Detection*	*Precise identification*	*Analysis*
Bridge	6	1.5	0.3
Radar site	3	0.3	0.04
Troop unit	6	1.2	0.08
Aircraft	4.6	0.9	0.03
Missile site	3	0.6	0.08
Ship	7.6	0.6	0.08
Port	30.5	6	0.3
Railway yard	30.5	6	0.6
Road	9	1.8	0.15

Source: *Reconnaissance Hand Book*, published by McDonnell Douglas; quoted in B. Jasani, ed., *Outer Space – A New Dimension of the Arms Race.*

from orbit, to 'watch' events on Earth of military importance. For example, the Soviets put into space at least three satellites of different types to monitor the early stages of the Iran–Iraq war in 1980. During the Falklands conflict in 1982, Soviet (and American) spy craft were flying over Britain's task force to watch out for anything interesting.

The second way of sending the information to Earth is by radio waves. The US tried this approach for the first time in 1961 with a surveillance vehicle called Samos 2. Spacecraft of this type have now been replaced in the USA by the aptly named Big Bird satellites – huge affairs that weigh 13 tonnes – and by the KH-11 vehicles. KH stands for Key Hole – military officials in the USA choose names for their space programmes which have a certain ghoulish appropriateness (we shall encounter other examples of this before the end of this chapter). Big Bird satellites are operated by the US Air Force; the KH craft by the Central Intelligence Agency.

The major space powers are turning to other kinds of sensors for mapping what is taking place on the ground. By the late 1980s, American armed forces will probably rely on radar instruments rather than ordinary cameras. These have the big advantage that they can 'see' through cloud. One type of American radar satellite will track the movement of tanks and other armoured vehicles operated by the Warsaw Pact, while the US Navy will use another to watch for warships.

An interesting development is the overlap that is taking place between military and civilian mapping satellites. As the latter increase in resolution, the information that they collect begins to find military applications. For example, Landsat 5, the USA's latest civilian remote-sensing satellite, can 'see' objects as small as 25 metres across. This provides good enough accuracy to make out, for example, places such as ports and railway yards that could become military targets. Spot, France's remote-sensing satellite that should be launched in 1985, will have a resolution of 10 metres, almost good enough to register missile silos. Third world nations, which have no access to military 'spy' satellites of their own, routinely buy Landsat data for use by military officers in preparing maps of nearby countries that could be used in any possible conflicts.

The fact that information from Landsat could be useful militarily was one reason why, in 1982, the US Department of Defense held up the sale to China by American companies of ground equipment that could tune in to transmissions from the satellite. Data from the satellite could help, for instance, to pinpoint troop movements on China's borders with the USSR and India. China has shown a lot of interest in satellite reconnaissance. Between 1975 and 1978, China put into orbit three photographic spy vehicles with its own rockets, presumably to provide information about possible targets for the country's nuclear missiles.

France is another country from the 'second league' of space powers that is interested in obtaining military information from the heavens. French military chiefs would like to launch a surveillance system called SAMRO. The technology would be based on France's civilian Spot satellite. The data from SAMRO would be helpful in guiding France's armoury of nuclear missiles to their targets, should they ever need to be fired. Budgetary cuts in 1982 forced the French government to defer the SAMRO programme, but it may get off the ground later in the 1980s. France is talking to the West German and British governments about a plan under which the three nations would jointly operate the surveillance system.

The links between the civilian and military uses of space were made plain by a delegation that called on Spot Image, the French company that will run the civilian remote-sensing service. The delegation consisted of military officials from Iran who told Spot Image officials that they wanted an Earth station to receive signals from Spot; the purpose was to pinpoint military installations belonging to Iran's enemies in the Middle East, Iraq for example. Spot Image politely but firmly told the military men: 'We are for civilian applications only.' But Iran, or any other country for that matter, could easily buy a Spot receiving station under the pretence of using it, for example, to monitor crop growth and then turn it to military applications later.

Britain, too, could well discover that a non-military remote-sensing craft could help it to defend one of the last relics of the empire. The European Space Agency's ERS-1 satellite will, from 1988, scan the oceans with radar instruments, mainly for scientific purposes. Britain is discussing whether to site a receiving aerial in the Antarctic region, where it has several research bases. Data from the vehicle would give scientists a better understanding of, for example, the build-up of ice or of tell-tale geological phenomena that could indicate mineral deposits. However, another aspect of a receiving station in the Antarctic will not be lost upon officials in Britain's Ministry of Defence, who must face the fact that at some point Britain may need to defend the Falklands against another Argentinian incursion. A satellite station in the region, perhaps even on the Falklands themselves, would provide highly accurate pictures of what is happening at sea. Certainly there would be no danger that Argentina could mount a surprise attack by sending warships in the direction of 'the Malvinas'. The vessels would be detected almost immediately by the radar vehicle; the news would appear on military chiefs' briefing documents shortly afterwards.

The reliance on satellites for reconnaissance does not mean that defence officials have lost faith in the surveillance methods practised during the Second World War. People watching the ground with a camera at the ready will continue to play a part in the way the USA and Soviet Union monitor each other's activities; but the super-powers will avoid, where at all possible, the use of aircraft. The diplomatic controversy caused in 1960, when the USSR shot down an American spy plane piloted by Gary Powers, taught the 'big two' to concentrate on space vehicles for manned surveillance. Thus the Salyut space stations put into orbit by the Soviet Union (see Chapter 7) routinely carry cameras for watching military activities on Earth.

The third and fifth Salyut stations, launched in 1974 and 1976, were devoted specifically to military operations. The USA, too, has a spy base in the sky in its fleet of space shuttles. About one-third of the shuttle's flights over the next few years will be on behalf of the US Department of Defense, and many will carry high-resolution cameras operated by astronauts recruited from the army, navy or air force.

The USA's permanent space station planned by NASA would do a similar job. The station is due to be built in the 1990s as a series of modules hoisted into space by the shuttle. According to defence analysts, a person in orbit watching for evidence of, say, nuclear missiles acts as a useful back-up to unmanned military satellites that may, on occasion, give false readings. The interest by the American armed forces in putting people in orbit marks a return to 1960s thinking. The US military wanted to go ahead with its own space base, the Manned Orbiting Laboratory, but this was cancelled in 1969 to cut costs. Military officials in Britain – which, with other West European nations, is to take a share in building the US station – have indicated that they would be interested in using the orbiting base for experiments in surveillance.

Photographic reconnaissance vehicles are often launched accompanied by other, smaller satellites nicknamed 'ferrets'. These are electronic devices that eavesdrop on radio communications. The craft record enemy signals on tape and replay the messages as the vehicles orbit over their own ground stations. Such ferrets played a part, we must assume, in the ability of US intelligence to monitor the conversations between ground control and the Soviet fighter that shot down the civilian South Korean airliner in September 1983 over the eastern part of the USSR.

One of the USA's earliest space projects in the military sphere was Midas, a network of satellites a few hundred kilometres above the Earth that was designed to give thirty minutes' warning of a Soviet missile attack. Midas satellites carried infrared sensors that were supposed to detect the heat given off by the exhausts of missiles as they cruise through space. The Midas programme has been replaced by similar early-warning satellites positioned in the geostationary orbit 36,000 km above the Equator. Vehicles that do this job for the USA include the Rhyolite series of craft and vehicles in what is called the Defense Support Program. One of the satellites in the latter is positioned over the Indian Ocean where it can monitor nuclear tests (as well as the launch of real missiles) in China and the Soviet Union. Another hovers over the American land mass, to sound the alert if submarines fire missiles.

The USA is continually trying to improve the accuracy of its infrared instruments to produce more sensitive early-warning satellites. American engineers are working on a programme called HALO (it stands for High Altitude, Large Optics, although this hardly gives away a great deal) which aims to produce a very sensitive infrared instrument with up to 10 million individual detectors. A similar project is called Teal Ruhy.

Another type of military satellite monitors the weather. In the USA, these vehicles are in a polar orbit in which the satellites travel in a line along the Earth's north–south axis. As the Earth is continually rotating, the vehicles will in this way pass over every segment of the planet's surface. The vehicles, part of the Defense Meteorological Satellite Program (DMSP), are similar to those that the National Oceanic and Atmospheric Administration, a civilian agency of the government, operates for non-military purposes. Data from the DMSP satellites can be fed directly to mobile terminals on ships or ground bases. They are also monitored at three large tracking stations: Fairchild Air Force Base, Washington; Loring Air Force Base, Maine; and another air-force facility in Hawaii. Information from all three locations is sent by satellite links to a central storage point at the Air Force Global Weather Control, at Offutt Air Force Base, Nebraska. Military meteorological vehicles can be crucial in providing information for forces, especially in places where weather conditions can change quickly. For example, in the Falklands war data from such craft were of profound importance, for instance in influencing the decision by British commanders about when to attempt to recapture the islands. It appears likely that, during the war, UK forces obtained such information via the DMSP satellites.

Geodetic satellites are a further variety of spacecraft that would be useful in war. These vehicles travel in a prescribed orbit around the Earth. Variations in this orbit are monitored by highly accurate tracking equipment on the ground. The variations are related to differences, from one place to another, in the gravitational force with which the Earth attracts satellites. These variations in gravity (which are due, for example, to the Earth being a less than perfect sphere and to differences in the density of rocks in the Earth's crust) will affect the path of missiles from their launch pads to their targets. If the weapons are to be completely accurate, therefore, the computers that control them must be given information about how gravity changes over the Earth's surface. The data from the geodetic vehicles can be fed straight into the computers' memory banks, increasing the probability that the missile makes a direct hit.

Navigation has always been important in warfare. To a soldier, the ability to know exactly where he is is nearly as important as knowing the whereabouts of his enemy. For centuries, people have navigated by the stars or with compasses, but over the past few decades navigating by radio beacons became possible. To obtain an indication of your whereabouts, you have to lock a radio transmitter onto a source of signals. With two radio sources, your position can be pinpointed in two dimensions.

The satellite equivalent of radio navigation has the title of the Global Positioning System or Navstar. This is an American military programme to build eighteen satellites orbiting in belts as high as 16,000 km. Receivers in land vehicles, back packs, ships, aircraft and even missiles will measure the time that radio signals take to arrive from different satellites. As the transmission times of the spacecraft are co-ordinated, the receiver can work out, with a small computer, its own position and speed. The system should be accurate in three dimensions to within a few metres. The Global Positioning System should be in full operation by 1988, when all the satellites should be in place (by June 1984, nine satellites had entered orbit). The system is an updating of an earlier set of American navigation satellites called Transit.

According to reports in the USA, the Soviet Union is considering whether to build a rival series of space hardware for navigation, under the name GLONASS (Global Navigation Satellite System). Though essentially for civilian use, for example by airlines, this could have military applications. The paradox about the Global Positioning System, however, is that in theory it has as much value to possible enemies of the USA as it does to the military forces that developed the system. Providing that the enemy knows the radio codes on which Navstar satellites transmit, there is nothing to stop it tuning in to the radio signals and obtaining its precise position. That is why the US Department of Defense is taking great care to keep secret the codes that will give the most highly detailed three-dimensional information. Civilian users of the Global Positioning System (of which there will be a large number, according to the plans of the hardware's developers) will be restricted to codes that give less precise information.

The final type of satellite of interest to the military is the kind discussed in some detail in Chapter 3 – communications vehicles. In essence, communications hardware for armed forces differs only slightly from that for the ordinary user of the telephone. Military communications craft carry digital information, pictures and the sound of conversations in much the same way as their civilian counterparts. But in one aspect the similarities end. Few civilians mind very much if they fail to get through on a transatlantic call because a transponder on a satellite is faulty; but such an event could be crucial on a military spacecraft in the middle of, or in the lead-up to, a war. That is why engineers go to great lengths to ensure that communications vehicles for military use do not fail or, if they do, that there is plenty of back-up hardware available.

This philosophy is manifested in the provisions taken by the US Air Force in the latest version of its Defense Satellite Communications System (DSCS). Spacecraft in this series are launched into the geostationary orbit in pairs (one acts as a support if the other fails). General Electric, the American company, is the contractor for the third generation of DSCS satellites.

The firm is shielding electronic circuitry in the craft so it withstands nuclear radiation. A nuclear detonation on the ground or in space would produce, as well as a huge blast and a lot of heat, a burst of radiation known as an electromagnetic pulse. This pulse would destroy circuits on board satellites. In the absence of electromagnetic shielding, satellites would become the first casualties of the early stages of a nuclear conflict.

In another piece of electronic ingenuity, technicians are ensuring that DSCS satellites' radio links are difficult to jam. If a ground-based jammer locks onto one of the satellite's radio channels, the channel shuts down and is replaced by one with different characteristics.

The US Department of Defense launched the first of the DSCS third-generation satellites (DSCS-III) in October 1982. Prior to this, it had put into orbit a total of eleven of the earlier-generation DSCS-II vehicles made by aerospace company TRW. (Launches are normally in pairs; two satellites being orbited by the same rocket. In addition to the successful launches, two attempts to orbit a total of four DSCS-II satellites have ended in failure.) The first pair of these satellites was launched into the geostationary ring in November 1971. According to plans for DSCS-III, General Electric will build seven of the vehicles for the US Military. Six of these will be positioned in four geostationary sites, over the Atlantic, Indian and east and west Pacific oceans. The final spacecraft is due to act as a ground spare.

In the late 1980s, the USA plans to put into orbit its next family of military communications satellites, codenamed Milstar. The vehicles will be built by Lockheed, under a $1,000 million contract. Milstar satellites will receive and transmit on frequencies that change perhaps thousands of times a second, a technique to foil attempts by enemies to listen to transmissions or to jam them. The system will be shared by all the military forces of the USA; different divisions of the army, air force and navy are developing the terminals (both mobile and fixed) with which military units will be equipped to communicate via the vehicles.

The chief feature of the Milstar craft is that they will transmit and receive on extremely high-frequency (EHF) channels in the 40 GHz region. (Most military satellites, in contrast, use the much lower-frequency 7–8 GHz channel, which is reserved by international convention for this type of traffic.) There are two reasons for this move to high frequencies. Channels in this region can carry more communications signals on the basis that the higher the frequency of a radio channel, the higher its bandwidth, or ability to carry information, becomes. Thus Milstar will be more suited to send between different parts of the world the very high volumes of data that, for instance, are required to prime and control modern weapons systems.

The second reason for the use of high frequencies is that Milstar is designed for use in a nuclear war. Satellite communications at around 40 GHz are thought to recover more quickly from the degradation caused by the explosion of nuclear bombs and the ensuing electromagnetic pulse. In the words of the US Air Force, Milstar is designed to be 'survivable during all levels of conflict'.

Other communications-satellite systems operated by the US Department of Defense include vehicles solely for the navy (this system goes under the name FLTSATCOM) and a set of hardware for the air force called AFSATCOM. The latter is not so much a series of satellites as an array of electronic equipment that is loaded, piggy-back fashion, onto communications vehicles operated by commercial undertakings. In this way, defence chiefs minimize the risk that all the air force's communications hardware could be destroyed in one fell swoop. In the USA, some 70 per cent of all the military long-distance communications traffic is routed by satellites; by 1990, the figure will be 90 per cent.

The most evident manifestation of the US military establishment's reliance on space technology was the setting-up in September 1982 of a new division of the US Air Force – the Space Command. The official motto of this new organization is 'Guardians of the High Frontier', a phrase that could have been borrowed from popular films of the *Star Wars* genre. In the hyperbole-laced words of the Space Command, the motto 'reflects the same spirit and determination that our pioneer forefathers had as they met the unknowns of a new continent, and as the Air Force meet the unknowns of the new frontier – Space'.

The Space Command is the leading agency for the US military space budget in 1984 of $9,000 million – a figure which has risen from $2,000 million in 1974. Some 6,000 air force officers and civilians work for the unit, which has four bases: Peterson Air Force Base, Colorado; Clear Air Force Station, Alaska; and Thule and Sondrestrom Air Bases in Greenland. The routine jobs of the command include maintenance and operation of the USA's military satellite systems which give information about meteorology and early warning of enemy missiles. The organization will also operate the Global Positioning System for navigation and the Milstar communications network. Space Command is responsible for the Space Defense Operation Center, based at Cheyenne Mountain, Colorado. This controls military satellites and receives intelligence, provided by observation satellites, about the military forces of possibly hostile powers. The base would also control any anti-satellite weapons that become operational. Also under the authority of the Space Command will be a new Consolidated Space Operations Center being built near the Peterson Air Force Base. When operational in 1986, this will have a staff of 2,000 and control military flights of the USA's space-shuttle fleet. Shuttles engaged on military missions will take off from and land at a new facility at Vandenberg Air Force Base near Los Angeles. This is to keep the payloads that the vehicles will be carrying, military communications satellites for example, well away from the civilian bases of NASA which will be responsible for commercial flights of the space shuttle.

The US Air Force's ground-based cruise missiles, some of which are stationed in European locations such as Greenham Common, England, are heavily dependent on satellite technology. If the missiles are ever fired, military communications links will play a major role. Three sets of connections link the launch crew for the missiles with commanders in both North America and the United States European Command near Stuttgart. If the weapons need to be deployed, the crews and their missiles (which have mobile launchers) will be in isolated positions, probably in woodland. It will be in such places that launch commanders await the order to fire. The first communications link is between the launch crew and the European headquarters. It uses the satellites in the AFSATCOM network and is known by the special code name 'Flaming Arrow'.

The second link connects the soldiers in their woodland clearing with military establishments in the USA and airborne command posts, which are converted Boeing 747 airliners. One of the elements in this network has the chilling code name 'Cemetery'.

The final radio connection uses land lines between the European headquarters and the main European base of the US Air Force at Ramstein, West Germany. From here messages can travel to the launch crews in the individual missile convoys. The proliferation of possible links is designed to ensure that, even if some connections are destroyed, the orders about whether or when to fire get through without any ambiguity. Once the missiles are on their way, other channels of communications would carry data between them and the ground, for example to update the guidance information in the weapons' computers.

Less is known about the Soviet Union's system of military communications, but it is a safe bet that the system relies on satellites similar in type to the USSR's civilian communications vehicles. These generally use non-geostationary orbits, so that people who tune in to the craft need to track accurately the satellites as they pass across the Earth at different points. (Being in fairly northerly latitudes, the Soviet Union lacks a suitable launch pad that can easily inject a satellite into the geostationary orbit over the Equator.)

As a country in the second division of military powers, Britain has less need for its own system of communications craft for defence. For keeping in touch with its forces over long distances, the country has generally relied upon either American craft or those operated by the North Atlantic Treaty Organization. These communications links would be needed, for example, for messages between the UK's fleet of Polaris submarines and headquarters in Britain.

Bolstering the radio channels provided by other people's satellites is a British geostationary satellite called Skynet 2B. Skynet 2B is, however, rather antiquated and cannot be relied upon to give a full service. It was launched in November 1974, and had an expected design life of only about seven years, although it has lasted longer. The Skynet 2A craft, launched in January 1974, failed to go into the correct orbit. (There were also two members of the Skynet 1 family; one was successfully injected into orbit in 1969, while a malfunction on the launch pad the following year accounted for the other.) Skynet 2B was too unreliable to be of much use during the Falklands war in maintaining contact between Britain and the country's armed forces some 8,000 km away in the South Atlantic. So the UK Ministry of Defence had to use an American military satellite (probably in the DSCS series) for communications links. Without this little-appreciated aspect of the Falklands campaign, Britain would have been forced to fight a modern war with communications links that were completely out of date. The episode frightened Britain's military boffins. They ordered that a new series of military craft, Skynet 4A and Skynet 4B, should go into service as quickly as possible. (Skynet 3 never saw the light of day; the Labour government cancelled it in the mid-1970s to save money.) The new generation of British craft should enter orbit in 1986 on separate space shuttle flights. British Aerospace and Marconi are building the vehicles in a joint programme.

The Skynet 4 satellites are similar in performance to the USA's DSCS-III communications vehicles. Both contain anti-jamming hardware and are intended to resist the effects of nuclear explosions.

For this reason, NATO is considering both designs for its new generation of communications vehicles due to enter orbit in the early 1990s. Up to the end of 1984, NATO had launched six communications satellites, all built by Ford Aerospace of the USA, which hover in the geostationary orbit above the Atlantic. Three of the craft are operational – they channel signals between ships and ground bases in NATO forces in Western Europe and the USA. If NATO chooses the British design, it will have been a coup for the UK manufacturers in breaking a US stranglehold on supplies of NATO space equipment. The two UK satellites are also due to enter orbit above the Atlantic. This marks a change in pattern, as the existing Skynet 2B craft went into orbit above the Indian Ocean – it was planned in the days when Britain had significant military forces east of Suez.

With all the importance that countries attach to their military satellites, it is only natural that top brains in defence laboratories around the world should try to work out ways of rendering the hardware ineffective. If one country could 'take out' a rival's space-based military system, it would be a step ahead in any outbreak of hostilities. Hence was born the antisatellite weapon.

The most obvious way of disabling an enemy spacecraft is to launch a second vehicle in the same orbit. With tried and tested manoeuvring techniques, the kind by which the USA and USSR routinely dock civilian space vessels, the second craft could come alongside the first. The 'aggressor' would contain a small load of explosive. A touch of a button at a ground control station and, flash, both vehicles disappear in a plume of smoke.

The approach does not require a great deal of advanced technology, so it is no surprise that at least one of the space powers appears to have perfected such a system. The Soviet Union has tested this kind of antisatellite weapon several times. On one recent occasion, in June 1982, a satellite called Cosmos 1379 was launched from the USSR's Tyuratum rocket base. Its purpose was to seek out a 'target', another spacecraft called 1375 that had left the launch pad twelve days earlier. After a brief period of manoeuvring, Cosmos 1379 came alongside the target. It was, with little doubt, a practice run for destroying an enemy space vehicle.

The USA is also interested in antisatellite techniques. From 1963 to 1975 the country had an operational antisatellite weapon which was based on Johnston Island in the Pacific Ocean. The system was based on a rocket booster which would home in on its quarry using infrared sensors. The hardware was dismantled, however, during the period of *détente* with the Soviet Union. At the same time the US Air Force's SAINT programme was terminated. This is an acronym for Satellite Inspector Technique, presumably a euphemism for a process of disabling enemy spacecraft.

With the renewal of cold war attitudes at around the time of the invasion of Afghanistan by Soviet troops in 1979, interest in the USA in antisatellite weapons has revived. The US Air Force is considering several possibilities. One of them follows a similar principle to the Soviet technique. An orbiting spacecraft would come alongside its target and blow it up. A more promising idea, however, is to launch what is called a homing-intercept vehicle from an F-15 aircraft. The vehicle would be like a small rocket and would climb rapidly from fairly low in the atmosphere to hit the enemy satellite. This technique has the advantage that to get the aircraft, and with it the interceptor vehicle, off the ground would take a matter of minutes. An antisatellite weapon of the Soviet design would take far longer to deploy because it is based on a rocket that needs many hours of preparation on a launch pad. Hence the USA, providing that its system works, would be far better placed to respond quickly to any emergency.

The USA can fairly claim that, unlike the Soviet Union, it has not actually fired its antisatellite system in earnest against a craft in space. In January 1984, an F-15 fighter fired the weapon in the skies above California, but it was not directed at a specific target.

Assuming that the US weapon is developed, it would have one other advantage over the USSR system. The latter can reach only those space vehicles in low orbits a few hundred kilometres above the Earth. The American weapon, in contrast, has the potential to hit satellites in the geostationary orbit at an altitude of 36,000 km. It is in this orbit where some of the most useful military spacecraft, for communications for instance, are positioned.

At what point would these antisatellite weapons be used? The most likely possibility is in the very early stages of a war. If either the USA or the USSR wanted to strike at the other, they would dearly love, first, to knock out key vehicles among the opposing country's military satellites. In the absence of such craft, a country would be at a serious disadvantage when it came to later phases of the fighting. The most frightening aspect of antisatellite weaponry is the part it could play in triggering off a nuclear war by accident. Increasing numbers of satellites are in the geostationary orbit. While the dangers of collision are mathematically remote (see Chapter 3), on several occasions in recent years near misses have occurred. For example, in 1980 an American early-warning vehicle and the communications satellite FLTSATCOM narrowly avoided hitting each other. Just a few days later, the same early-warning satellite nearly made contact with a civilian communications craft operated by Western Union.

Suppose such an accidental collision involved a sensitive military craft. One moment the satellite is present on space controllers' radar screens, and the next it vanishes. What would the satellite's owner make of this? Would it assume that the 'other side' had destroyed the vehicle as a prelude to blasting a salvo of missiles and starting a war? It could then be all too easy for the country that feels it is on the receiving end to swing into action with its own missiles. Possibly no one would ever know a mistake had been made – or if they did, it would be only after much of the Western world had become a nuclear desert.

Another way in which satellites could be disposed of is with hardware that in the USA and USSR is still on the drawing board – beam weapons, rays of high-energy particles of light. These armaments have become the most discussed aspect of the superpowers' strategies for fighting a war in space. This is in spite of the fact that the installation of such weaponry, if it takes place at all, will come about probably no earlier than the early part of next century.

Beam hardware could function not only as antisatellite, but also as antimissile, weapons. The latter are designed to knock out intercontinental missiles as they travel through space and before they reach their targets. Antimissile weapons based on beam technology could form a defensive screen, of the kind envisaged by American scientists and military officials working on the 'Star Wars' programme, around cities, or even whole countries.

Antimissile weapons are far from new. For a decade or more, engineers have been developing rockets and spacecraft that home in on other missiles and destroy them. Such hardware is used, for example, to counter short-range rockets (which travel only a few tens of kilometres) fired by aeroplanes and ships and which feature in small-scale, non-nuclear conflicts such as the Falklands war. Higher up in the scale of military technology are the antimissile systems that both the USA and USSR have developed to defend themselves against long-range intercontinental missiles. Such 'first-generation' systems, in the terminology of defence technologists, are based on kinetic-energy devices, in other words ordinary rockets. Both the USA and the USSR developed such antimissile hardware in the early 1970s. In the mood of *détente* that then prevailed, the superpowers agreed in 1972 to an Anti-Ballistic Missile Treaty that limited the number and geographical coverage of such systems. The feeling was that if one country installed too many defensive screens, destabilization would occur on the grounds that the more technically advanced nation might be tempted to launch a first strike without risking retribution from the adversary.

The Star Wars programme, as outlined by President Reagan early in 1983 and fleshed out by US engineers and defence officials later in the year, adds a new dimension to discussion of antiballistic-missile technology. There are two reasons for this. First, the system would feature not just kinetic-energy projectiles but 'directed-energy' beams either of particles such as neutrons or of light. The latter would be in the form of laser radiation – or high-energy, highly coherent photons that can be focused accurately at a target. A beam weapon has as its main advantage over a projectile the fact that it directs its energy at the target in a much shorter time. A missile that uses conventional chemical-propulsion techniques travels in the order of a few kilometres a second; a beam of charged particles may have a velocity of a few thousand kilometres a second; while a laser beam travels at the speed of light, or 300,000 km a second.

While, therefore, a conventional ballistic missile would take around half an hour to voyage a quarter of the way round the world (roughly the distance between the USA and the USSR), a laser beam could travel the same distance in a fraction of a second. Thus it becomes possible to envisage a system in which a high-energy light ray from a battle station in one country shoots down, in an instant, a missile, newly launched by an adversary, that is just starting its cruise through space. A set of hardware reducing response times by such a degree would radically change the theories for fighting wars.

The second reason for the shock-wave effect of the USA's Star

Wars programme is that, for the first time, it envisages in detail the set of extraordinarily complex technologies needed to make an antimissile system virtually fool-proof. What is more, the programme started by President Reagan has signalled the USA's intention, no matter how impossible it may seem, to operate such a system by early next century. The difficulties are enormous. The antiballistic-missile weapons, which could be based either on space stations or on the ground, would have to deal with perhaps 1,000 intercontinental rockets launched from the USSR, each of them producing ten separate nuclear warheads. That makes a total of 10,000 objects travelling through space (not to mention 'decoy' warheads that might be launched simultaneously) which would somehow have to be detected and destroyed. If just one of the projectiles slipped through the screen, the result would be the impact, somewhere in America, of a nuclear device with several times the blast power of the Hiroshima bomb. This would mean that the defensive exercise had failed.

The USA's strategy, therefore, involves several layers of defence technologies, using a mixture of lasers, particle beams and 'first-generation' antimissile missiles, which would destroy incoming warheads in phases. In this way, if an enemy projectile slipped through the first layer, it would stand a good chance of being knocked out by the second layer; any stragglers managing to avoid destruction in this phase would be likely to be picked off in the next layer; and so on.

There are four phases in the way that intercontinental missiles fly to their targets. In the initial boost phase, which may last five minutes, the vehicle blasts off from its launch pad to reach an altitude of more than 100 km. In the deployment phase, a manoeuvring operation takes place in which the single missile splits up into perhaps ten individual warheads each of which is separately targeted. In the midcourse stage that follows, the projectiles move through space in a ballistic trajectory en route to the target. This takes perhaps twenty minutes. Finally comes the re-entry phase in which the warheads dive back through the atmosphere in a matter of one or two minutes to impact on the hostile territory.

From the point of view of defence planners, it makes sense to try to destroy enemy rockets in the boost phase, before they have split into independent warheads. This is where the first layer of a Star

Wars defence would aim to cripple perhaps 90 per cent of projectiles in the sky. Later layers would aim at intercepting opposing missiles in the other phases of their trajectories. The use of beam weapons would be especially valuable in this first layer of defence. The speed of operation of these weapons would be amazingly high – they would be fired as soon as the USA had detected the lift-off of Soviet missiles and would have a good chance of hitting the projectiles while they are still in the boost phase of flight.

Research work on lasers is strongly emphasized in the Star Wars project. In a laser, the atoms of a certain medium (which could be a gas, solid or a mixture of chemicals) are forced, or pumped, into an excited energy state. The atoms must return, at some point, to their original energy state, in the process giving up their energy by emitting electromagnetic radiation. This can be ordinary light or higher frequency waves such as ultraviolet radiation or X-rays. Under certain circumstances, what is known as 'population inversion' occurs, in which more of the atoms are in the excited state than at the lower energy level. If this occurs, the 'pumped' medium may emit a stream of monochromatic radiation that is known as a laser beam.

Lasers were invented only in 1960. Since then physicists have considerably increased the power output of lasers. According to some scientists in the USA, power levels of up to several million megawatts may be attainable. At this kind of power, lasers would find applications as an offensive weapon, so long as a means could be found of aiming them accurately. In terms of physics, such a laser device would be a very good way of transferring a large amount of energy accurately from one place to another, with possibly devastating results.

Defence establishments in the USA have already tested high-energy lasers in several experiments. In 1973 the air force used a carbon dioxide gas-dynamic laser to shoot down a winged drone at an air base in New Mexico. The army and navy have also demonstrated that laser weapons will work.

One of the most powerful lasers known to have operated is a 2.2 million watt device called MIRACL (Mid Infrared Advanced Chemical Laser). This is used at the USA's High Energy Laser National Test Range in New Mexico. A large chemical laser called Alpha, under construction for the Defense Advanced Research Projects Agency, should produce 5 million watts.

Particle-beam weapons, in theory, follow the same principle as laser hardware. The idea with these is to shoot a stream of high-energy atomic particles such as protons or electrons at a target. These particles would need first to be accelerated to a high energy by a series of powerful magnets in equipment similar to that used by researchers in high-energy physics.

There are two types of particle beams produced by high-energy accelerators, charged and neutral. Charged beams contain streams of electrons or protons; neutral beams use uncharged particles such as hydrogen atoms. In the USA, the navy is working on charged beams in a programme called Chair Heritage which mainly applies to short-range weapons for defending ships. Neutral particle beams are judged to be better for space applications because they are not distorted or deflected by the Earth's magnetic field. The Defense Advanced Projects Research Agency is concentrating on this kind of beam with a project called White Horse.

According to estimates, the USA is spending about $200 million per year on research in laser weapons and about a fifth of this total on particle beams. The Soviet Union probably also has similar programmes – though they almost certainly fall far short of Star Wars – but about which relatively little is known.

Laser weapons are in a more advanced state of research, and so any beam system for space of the kind that the USA is considering is more likely to be based on this principle. There are considerable technical problems, however. First, any high-energy laser system requires a power source that is big and very heavy. So a space-based laser battle station would be enormous, probably with a mass of several hundred tonnes. This would hardly be a secure item of military equipment, bearing in mind that conventional antisatellite weapons would find it an easy target.

This problem could be avoided by siting the laser on the ground, perhaps on top of a mountain. It would beam radiation to an orbiting mirror that would reflect the 'death rays' onto their targets, whether these are enemy satellites or incoming ballistic missiles. An alternative would be to deploy what is called a 'pop-up' system, in which the laser equipment is mounted on rockets that are launched only if an international crisis seems imminent. In this way the battle station stays in space, where it poses a massive security risk, for only the shortest possible time.

A second major difficulty is to ensure that a laser system can be aimed accurately. Laser beams are distorted as they pass through the atmosphere, so any ground-based laser weapon, which has to shoot radiation at the speed of light to an orbital mirror or to the target itself, would require sophisticated electronic compensating mechanisms to allow for this inbuilt margin for inaccuracy. A ground-based laser could also quickly become inoperable as a result of storms or other freak weather conditions which would 'blind' the optical instruments with which it tracks enemy projectiles.

Even if the laser system were based on a battle station in space, where distortion is less acute, ensuring accuracy would still be a problem; there would be very little margin for error. This poses a gigantic headache for defence planners. They would require extremely advanced computers, employing principles of artificial intelligence, or the ability to operate in a similar way to the human brain. The computers would first track the missiles and then direct the laser at the right target. Department of Defense programmes known as the Large Optical Demonstration Experiment (LODE) and, still more cryptically, Talon Gold are aimed at tackling some of these command-and-control issues.

A test which illustrated how these problems are being tackled occasioned world-wide publicity in June 1984. A rocket launched by the US Army from a base in the Kwajalein Atoll in the Pacific intercepted and destroyed a rocket that had lifted off minutes previously from California. An infrared sensor on the 'killer' rocket tracked and located the Minuteman missile at an altitude of about 160 km, at about the time it was starting its re-entry phase. Shortly before impact, the warhead on the interceptor unfolded like an umbrella to deploy a steel net about 5 metres in diameter. This engaged with the incoming rocket, destroying it in a simple collision.

The main object of the exercise was to try out the sensor, which homed in on the exhaust of the Minuteman rocket and guided the army missile to the point of contact. The difficulties of effecting the collision, when the bodies were travelling with a velocity relative to each other of about 6 km a second, or around six times the speed of a bullet, should not be underestimated. However, this was a very straightforward trial, by the standards that the USA's military planners have set themselves. The weapon deployed was a conventional first-generation 'kinetic energy' device. What is more, there was no attempt to destroy the Minuteman during the early boost phase when, in real fighting, engagement and destruction would be that much more valuable.

The most detailed look at how the new generation of antimissile technology could work came in a report towards the end of 1983 by a military group called the Defensive Technologies Studies Team. The group, headed by James Fletcher, a former administrator of NASA, outlined to President Reagan the technical challenge of building the system he had talked about in his Star Wars speech the previous March. Research on the hardware, continuing where possible from existing Department of Defense programmes on lasers and other technologies, would cost between $18,000 million and $27,000 million from 1985 to the end of the decade. A fully operational antimissile system could enter service during the 1990s at a cost of around $95,000 million.

The report, parts of which were published in the American magazine *Aviation Week and Space Technology*, envisages hardware based on Earth and in space that would identify, track and then destroy enemy missiles. The hardware would include chemical lasers, neutral particle-beam weapons, airborne optical 'spy' equipment and X-ray lasers. The latter produce high-energy X-rays and are powered by the detonation of a small nuclear bomb. The whole array of armoury would be co-ordinated by a 'surveillance and battle management' station high in orbit about 100,000 km above the ground.

The Fletcher report claimed that to work on the programme in the way envisaged would help the USA in relations with the USSR. 'Even prior to development, the demonstration of US technology would strengthen military and negotiating stances, and options for immediate deployment would play a significant role in deterrence.' This philosophy is based on the notion that, if the Soviet Union is faced with a system that renders US cities safe from attack, it would be less likely to threaten aggression. It would also soften its line on matters such as arms control and human rights. An indication that the USA intends to use its Star Wars programme in this way, as a kind of bargaining chip to gain concessions from the USSR in other areas, came in January 1985. At a two-day set of talks on arms issues between the two countries in Geneva, they agreed to new discussions at which, so observers speculated, the USA could promise not to deploy its Star Wars system (assuming it can be perfected) in exchange for positive moves on arms control from the Soviet Union.

Opponents of the Star Wars initiative, on the other hand, say that the USSR may grow, as a result of the programme, not more submissive but more belligerent. The USSR could be propelled by President Reagan's plans either to redouble its own military research and development to find ways of pushing its rockets through the envisaged defensive net or, still worse, to do something unpredictable that could spark off a nuclear conflict well before the Pentagon's scientists and engineers had time to demonstrate their technical expertise with a protective system.

Some critics have pointed out that the technological leaps which the system would require look unrealizable. Professor Hans Bethe of Cornell University, and a former head of the theoretical physics division at the Los Alamos Atomic Scientific Laboratory, made his feelings known in the March 1984 edition of the US magazine *Laser Focus*. He said that the envisaged laser system would probably not work and that the intensified arms race 'would increase suspicion and therefore . . . really make war more likely'. The computing problem by itself would be daunting, Professor Bethe declared:

> You need computers far beyond the state of the art. Your battle management stations must keep track of over a thousand boosters. Of course, if there are such space defenses, the Russians would probably launch all their boosters within a few minutes. You would have to keep track of a thousand laser stations [each of them targeted on a particular booster]. You would have to assign targets to each laser. You should try to ascertain whether a given booster has been destroyed or not. That seems to be one of the most difficult problems.

A swingeing critique on laser weapons, delivered on both political and technical grounds, came well before the Star Wars programme was officially instigated. In a report in 1980 two scientists at the Massachusetts Institute of Technology, K. Tsipis and M. Callaham, made several observations about weapons systems in general:

> The development of new weapons systems in the US since World War Two has been characterised by the 'fallacy of the last move'. This is the ever unfulfilled expectation that recourse to a new, and ever more exotic weapon system will either confer absolute military superiority over the Soviet

Union or will be the 'quick fix' to cure a problem that may not admit technical or military solutions at all because it is essentially political, economic, diplomatic or a combination of all three.

The scientists sum up on the subject of lasers in space:

> A laser spaceborne ballistic missile defence system would generate profound strategic instabilities since one could not be certain that, once complete, it would not be both effective and invulnerable. Since in all probability such an ideal system cannot be built, it is not worth attempting since, at worst, it will precipitate an armed conflict between the US and the USSR and, at best, it may induce an increase in the number of USSR launchers limited only by cost.

Turning to the idea of stationing laser weapons on the ground or in the atmosphere, the MIT workers say that such hardware would, at least, discourage other countries from putting battle stations in orbit which could then be shot down from beneath.

> On the other hand, any laser anti-satellite system that could suddenly decapitate the opponent's defensive structure, i.e. deprive it of its spaceborne 'eyes', 'ears' and command, control and communication facilities, would create a profound crisis instability, especially if it were mirrored by other nations.

The array of spaceborne high-technology apparatus on which the superpowers are working – ever more accurate missiles, satellites of increasing sophistication and now laser battle stations – form the backcloth to the changing strategies that military analysts are proposing for the way in which a nuclear war would be fought. The old theory about nuclear conflict is based on MAD – mutual assured destruction. According to MAD, a nuclear war is virtually unthinkable. In the event that such a conflict started, both the USA and the USSR would set off their missiles at each others' cities and industrial complexes, killing tens of millions of people and rendering huge tracts of the planet uninhabitable.

According to this thinking, nuclear war could start only as an accident or by the actions of someone who is insane. This thought is, in a strange way, comforting. It narrows enormously the ways in which a nuclear war could start. Thus efforts to ensure that such a thing never happens are reduced to a level that is at least manageable. Anything that supports the theory of MAD, which means the hefty arsenals of bombs ready and waiting to turn two of the world's biggest and most important countries into wastelands, is therefore to be welcomed.

However, this straightforward view of nuclear hostilities is under attack. In its place are theories which propose a whole range of scenarios in which nuclear war would take place. There would be several 'stages' to the conflict. First, precision-guided missiles would 'take out' enemy weapons installations, followed by key targets such as ports, airfields and factories. Satellites would help enormously, both in pinpointing the targets and guiding the missiles to their landing points. Only in later phases of the hostilities would the adversaries invoke the 'big stick' in which huge stocks of nuclear armaments are detonated over cities. However, such a stage may never come if one side gains so much of an upper hand early on that the opposition is forced to surrender.

In other visions of how a world conflict would develop, antisatellite hardware would disable important key space vehicles belonging to the enemy, so weakening vitally its military infrastructure. A country could, perhaps, justify this action on the grounds that a specific enemy satellite, or set of satellites, constitutes a danger to its own security and so must be destroyed. The same nation may be prepared to take this obviously aggressive action because it believes itself to be in no danger. This may be because the country's military boffins have erected an antiballistic missile defence system which, so they believe, will keep out any rocket attacks that the other side is provoked into making.

Other theories of nuclear warfare confine the conflict to small areas of the world, such as Europe, or even individual countries within Europe. This approach would minimize the dangers to the populations of the USA and the USSR, the countries which are likely to be the biggest protagonists in any dispute.

Underlying the ideas about nuclear war that have replaced MAD is the concept that a 'limited' nuclear war is conceivable and, still more to the point, winnable. So at some stage, if it believes the stakes are high enough, a government may be attracted to the notion of deliberately provoking an outbreak of hostilities. The network of space-based hardware on which it has spent so many billions of dollars (or roubles) may encourage the government that this course of action can be justified.

There is one problem with all this theorizing. It may be completely wrong. To start with, the wonderful technology may fail to work. For instance, the communications links may go dead or the anti-missile system crack up. Once a nuclear war starts, it may go totally out of control. The escalation from a 'small-scale' conflict to a full exchange of intercontinental weapons may take place too rapidly for anyone to stop it. Paradoxically, the superpowers' dependence on space-based links and control systems, far from providing the conflict with some semblance of order, could well hasten the slide to Armageddon. In the early stages of hostilities, each side's technological command-and-control apparatus may be destroyed, leaving the leaders of the war-mongering countries growing more confused and harassed by the minute, with little alternative but to press the relevant buttons and condemn the world to unbelievable chaos and misery.

The world contains some 50,000 nuclear warheads. They have a net destructive power of some 20,000 megatons of TNT, roughly equivalent to one million Hiroshima bombs. It is difficult to resist the conclusion that, to weaken the likelihood that the world will some day unleash at least a proportion of this firepower, the superpowers ought seriously to consider dismantling some of the military apparatus they have put into space.

7 ☆ Space Work

Products made above the atmosphere will go on sale in the early years of the next century. They will be manufactured in large orbiting complexes which capitalize on the special characteristics of space – low gravity, vacuum, relative freedom from impurities and abundant energy easily obtainable by turning the Sun's rays into electricity with solar cells. After the initial razzmatazz, people will think it no more unusual to buy goods stamped 'Made in Space' than they do now to buy those produced in the Philippines or Taiwan.

The trends that will bring about industrial parks in space are already well established. The Soviet Union is ahead in this area. It has put into orbit seven small space stations, the last two of some sophistication. Such a station can dock with other spacecraft to produce an orbiting complex up to 50 metres long, an outpost in the heavens the length of Nelson's Column. The USSR's Salyut space bases can hold, at a squeeze, five cosmonauts at a time. People have lived and worked in the stations for up to eight months in one stretch. The Soviet Union is keen to develop new, much bigger orbital complexes and site in them small factories that produce goods difficult to make on Earth.

Meanwhile, the USA plans to build in the 1990s a space station that would emulate the Soviet efforts. The $8,000 million base, in which people would live permanently, would be built in modules taken into orbit by the space shuttle. Western Europe, under the auspices of the European Space Agency, may decide to help. The work would follow from the agency's development of Spacelab, the world's first re-usable orbiting laboratory, which made its maiden mission, on board a US space shuttle, at the end of 1983. Spacelab promises to provide a 'shirt-sleeves' environment in which scientists can do experiments that may lead to new industries above the atmosphere in the next century. The laboratory, however, stays in space for no more than ten days or so at a time inside a shuttle's cargo bay, whereas a space station, with its own propulsion system and life-support equipment, may have a life of several years.

Space stations are useful to industrialists and scientists primarily because of the exotic conditions that space has to offer. In a spacecraft in a state of free fall in orbit around the Earth, strange things happen – or, to be more precise, don't happen. The gravitational force experienced by people or objects inside the spacecraft is between one-thousandth and one-millionth of the figure on Earth. As a result, many of the physical effects that we have grown used to on our planet are simply non-existent.

For example, substances will mix far more evenly because their weight is no longer relevant in determining how they diffuse when shaken up with other liquids or solids. Thus in space you would have relatively little trouble ensuring that, for instance, sugar is stirred into your tea. A bartender mixing drinks in the cocktail lounge of a space station could do his job with a lot less effort than on Earth.

Other familiar events would no longer occur, for example the sedimentation of particles in the oil in the gear box of an engine. Instead, the particles would be distributed uniformly throughout the liquid. In a similar way, it would be supremely simple to make fruit jellies in which the lumps of pineapple, pear and so on are evenly dispersed. They would not follow the annoying trend in Earth-bound kitchens and end up on the bottom of the container.

As a result of the absence of what is called hydrostatic pressure, viscous materials would not tend to collapse under their own weight. Hence slowly solidifying castings would be no wider at the base than at the top. And anyone who lights a candle in space would find that the steady trickling-down the stem of oil, a source of mild concern to any Earth being who has suffered a power cut, would be a legacy left behind on the launch pad.

In manufacturing terms, the near absence of gravity holds out remarkable prospects. It may be possible in space stations to make a bewildering array of materials difficult, or downright impossible, to produce on Earth. Researchers in the USA, USSR, Japan and Western Europe have drawn up a list of the materials in this category. They include alloys, superconductors, new types of glass, semiconductors, detectors for infrared light and nuclear radiation, catalysts, enzymes, polymers, superplastics, magnetic materials, corrosion-proof coatings, solar cells and high-strength composites or special steels.

The specific processes that are important, according to the researchers, include crystal growth, casting of metals, containerless processing, chemical reactions (e.g. polymerization), and biological separation through mechanisms such as electrophoresis, in which materials are isolated from mixtures according to the different electric charges on their constituent particles.

In addition to these areas of technology, any process that requires a lot of energy may be suitable for space factories on account of the abundance of energy available from the Sun. By the time the radiation from this source reaches ground level, much of its power is absorbed by the molecules in the atmosphere. Above the mantle of gas that separates us from the rest of the Universe, it would be an easy matter, using solar cells, to turn the full power of the Sun into electricity to drive industrial processes. Among the energy-intensive operations that scientists think may take place in space factories are metals smelting, welding and cutting with electron beams, joining up pieces of metal by brazing, and the sintering of powders to produce solids.

Of all the disciplines that appear promising for space industrialization, crystal growth holds out perhaps the most exciting possibilities. On Earth, scientists generally make crystals either by separating them from a solution or by techniques of vapour deposition. In the latter, a solid lattice is 'grown' by depositing gas onto a cool surface. Gravity interferes with both methods. Because the different parts of the crystal have a specific weight, the crystal can bend or sag, in the same way that a building suffers from stresses arising from different loads. Even the minute deformations that result from these forces may be enough to cause cracks or other structural defects that can change the performance of the crystal.

In the absence of gravity, however, it should be possible to produce crystals that are near-perfect. Thus space factories may turn out semiconductors such as gallium arsenide which have much better electronic properties than anything obtainable on Earth. Alternatively, semiconductor crystals may be grown with ease to an enormous size. In conventional factories, electronic circuits made from silicon are normally produced from circular 'wafers' no more than 10 cm in diameter because of the difficulty of producing pure silicon in larger dimensions.

Other crystal products could have particularly good magnetic or optical properties; materials such as compounds of mercury, tellurium and cadmium that are useful as infrared detectors could be made. Scientists also believe that defect-free crystalline substances made from gallium and bismuth, for example, could turn out to be good superconductors.

The company that has shown, in public at least, the strongest enthusiasm for growing crystals in space is Microgravity Research Associates, based in Florida. The company has signed an agreement with NASA to co-operate in experiments on shuttle flights. Bob Pace, Vice-President of Microgravity Research Associates, says he hopes, by 1990, to produce 40 kg a year of gallium arsenide made in space. The material would be sold to electronics companies requiring particularly pure specimens of the substance for ultra-efficient microchips. Companies in this position would have to pay a premium for material manufactured in this way. Mr Pace plans to sell his gallium arsenide at $1 million a kilogram, roughly 100 times the 1984 price of gold. Gallium arsenide made by conventional terrestrial means costs $20,000 a kilogram. By 1997, says Mr Pace, annual production would increase to 1,200 kg. By the end of the 1990s, the figure would reach 3,000 kg, at which point the price would have dropped to $500,000 a kg.

Solid objects cast from molten metals or plastics in Earth-bound forges often suffer from imperfections. These are due to particles migrating through the fluid as a result of convection currents induced by gravity. Such currents are a particular problem in composite materials where substances in a liquid form are solidified around fibres, of carbon or plastic for example. Alternatively, liquid metals can be made to harden around particles of, for instance, tiny spheres of aluminium oxide (alumina). In both cases, the extra materials lend strength to the finished product, but the effect can be ruined if the structure contains imperfections.

The presence of convection currents due to gravity may mean that some substances do not mix at all, or do so only under special conditions. This may preclude scientists from making castings or alloys from materials that, in theory, would produce unusual properties, for example in terms of strength, when combined. One example is the difficulty of making compounds in Earth-bound laboratories from gold and germanium; lead, tin and antimony; or aluminium and antimony. In space, researchers would have virtually *carte blanche* to try their hand at mixing substances irrespective of what can or cannot be done in terrestrial workshops. Hence a new range of alloys and compound materials could become possible, with applications in many industries from consumer goods to vehicles to construction.

John Deere, the US manufacturer of farm equipment, is another company that intends space experiments. It is interested in testing how carbon and iron mix in zero gravity. The interreaction is a crucial one in ironmaking and, with gravity 'turned off', the company may be able to gain a new understanding of the mechanism which would help in its operation of terrestrial iron foundries.

In the third broad area of low-gravity techniques in which researchers are interested, factory workers should be able to handle liquids without containers. In space, liquids have no weight and so are held together as spheres by surface tension. One result is that processes to make highly sensitive chemicals that can be harmed by impurities will make ideal candidates for manufacturing plants in the heavens. Chemicals in this category include certain kinds of catalysts. Such reactions may be difficult in ordinary chemical plants because the containers in which they take place introduce contaminants. In space, the liquids in the reaction would float around surrounded by nothingness. They would be held in place long enough for chemical or physical reactions to occur, by electrical or magnetic fields.

A specific application for containerless processing is the manufacture of glass. The material present in many millions of window frames around the world is, in reality, rather special. Glass is best described as a supercooled liquid. The atoms in it are frozen in position very quickly, before they have time to arrange themselves in the regular lattice-type structure of a solid. This is the basic explanation for the specific optical and structural properties of glass. However, although the manufacture of some types of glass, the sort made from silicon for example, is fairly easy, glass made of other substances may be very hard to produce. With these materials, the sudden cooling from the liquid phase is difficult to achieve. One way around the problem could be to cool the liquid in space without a container. On Earth, the receptacles in which the cooling takes place inhibit the ability to remove heat quickly, as they act as a point of nucleation around which the liquid can start to solidify, nullifying the whole idea of producing glass. Containerless processing could, then, lead to new kinds of glass with unusual properties, for example for use as lasers or to form a conduit for electrical signals in strands of optical fibre.

Polymer chemists are also interested in experiments in the heavens. With no gravitation to worry about, researchers may be able to build long chains or molecules of a kind that, on Earth, would be too unwieldy or which would snap under their own weight. In particular, scientists think they could produce new kinds of polymers known as latex spheres, which are tiny balls of polyvinyl latex. Latex spheres formed in the same reaction are of near-identical dimensions – of the order of a few micrometres – and so can be used as a calibration standard in microscopy. In space, researchers think they will be able to vary the dimensions of these balls rather more easily than on Earth, for example to build up bigger ones than are possible in conventional laboratories.

The final area that is, according to scientists, ripe for exploitation in orbiting complexes concerns a way of separating chemicals from mixtures in which they are naturally present. Electrophoresis is a technique tried-and-tested on Earth. An electric field channels substances in the mixture into different compartments depending on their electric charge. With this technique, workers can produce high concentrations of specific chemicals from, for example, human or animal secretions in which the substances are naturally present. For many important substances, though, the process is painfully slow, and as a result it is often uneconomic for ordinary factories. In space, however, the lack of gravitational force speeds up the technique possibly by several hundred times, depending on the chemicals involved. Substances that could be produced in spaceborne electrophoresis laboratories include many materials useful in medicine. Examples are beta cells, useful in treating diabetes; interferon, a drug that may be able to treat cancer among other diseases; growth hormones, for stimulating bone development in children; urokinase, which can prevent blood clots; erythropoietin, for kidney failure; and epidermal growth factor, used to heal burns.

Space electrophoresis is probably closer to commercialization than any of the other ideas for orbiting factories. McDonnell Douglas and Ortho Pharmaceuticals (a division of Johnson and Johnson) say they want to put into orbit, by 1988, a prototype electrophoresis facility. This would probably be on an unmanned platform and would be lifted into space, perhaps 300 km above the Earth, by the space shuttle. The two companies have already tried several experiments in space with small-scale electrophoresis hardware, the first of which was on the space shuttle's fourth flight in June 1982.

Other North American firms that have expressed interest in capitalizing on specific aspects of manufacturing above the atmosphere are TRW (production of turbine blades); GTI (castings and metallurgical furnaces); Du Pont (catalysts); Battelle (biotechnology, for example the growth of collagens); Westech Systems (large silicon crystals); Boeing (pharmaceutical products); General Electric (latex spheres); 3M (new materials); Kodak (fluid physics and production of new chemicals for film processing); and Inco, a Canadian nickel producer (electroplating techniques). These firms have either signed agreements with NASA to operate experiments in space over the next few years or are doing research in terrestrial laboratories to investigate the prospects for space manufacturing.

The industrial interest in making things in space is probably as great in the Soviet Union, which is behind the West in many conventional manufacturing activities and so has still more reason than the USA to start processing operations above the atmosphere. With the know-how and the products so obtained, the USSR could compensate for, at least partially, its deficiencies in other spheres of industry. The Soviets have undoubtedly done more practical work on this subject than the USA in their succession of space stations. Concrete details of what the USSR is planning, however, seldom leak out.

Most of the important Soviet activity in space experiments has been on Salyut 6 and Salyut 7, launched in 1977 and 1982. The USA has also made strides in materials processing in studies in their first (and so far only) space station, Skylab, to which astronauts were sent in 1973. Spacefarers from both countries have also carried out isolated experiments in short trips in space capsules. Such studies in weightless conditions have supplemented the experiments that, for twenty years, have taken place with sounding rockets and aeroplanes. When such a vehicle accelerates to the top of a ballistic trajectory and then suddenly dives, the physical forces acting on the craft are such that a brief period of weightlessness is obtained. But such spells last only a few minutes at the most and are not much use to researchers who are serious about convincing people of the merits of spaceborne processing. It is safe to assume that we will see many more advances in this area of technology in the new, more sophisticated, orbiting platforms that should be launched over the next few years. For a couple of decades, the work will stay firmly in the research phase, leading to fully-fledged space factories some time after the year 2000.

A permanent platform in space has attractions not just to the hard-nosed industrialist. Scientists engaged in the purest, most non-commercial forms of research are interested in space laboratories. They think that with gravity 'switched off' they may obtain fresh discoveries about a great range of phenomena.

Ever since men and women started systematically to study their surroundings, so the argument runs, they have had to contend with gravity. It is an ever-present force that continually 'modulates' scientific results. If, over the past couple of hundred years, scientists had been able to alter gravity as easily as they can modify conditions of temperature and pressure, then our understanding of the physical and biological world would probably be enormously improved.

In the near absence of gravitational attraction in space laboratories, scientists think they will be able to shed new light on many of the basic properties that affect materials, how substances conduct electricity or how their molecules link together, for example. Biologists are also interested in low-gravity work. In workshops above the atmosphere, they may be able to investigate how human cells grow and evolve, or the mechanisms that affect bodily stress and heart disease. Experiments in space could produce an understanding of the nature of the gravity-sensing organs of the body, for example the vestibular apparatus of the inner ear. Experiments may also be devised to examine the basic questions of how people and animals have evolved to cope with gravity, for example the way bone structures grow.

An indication of the importance of medical studies in space came with the results of the first Spacelab mission in 1983. Scientists monitoring the astronauts on the flight think that, for the first time, they have a real understanding of the causes of 'space adaptation syndrome' – an illness which produces sickness and nausea and affects many space travellers during their first few days of weightlessness. Researchers think they have pinned down the cause of the ailment to the otolith, an organ inside the inner ear that sends electrical signals along nerves to the brain and so helps people to keep their sense of balance. In space, according to the Spacelab results, for some reason the behaviour of the otolith is disrupted. It starts to send 'rogue' information to the brain which contradicts the information coming from the eyes. Not surprisingly, the brain becomes confused and the sickness is the result.

Astronauts, it appears, differ in the degree to which they suffer from the nausea, depending on the way the otolith interacts with other parts of the body under normal conditions of gravity. This interaction can be quantified by monitoring the signals to and from the otolith before the astronaut takes to space. In this way, it may be possible to predict those most prone to the syndrome and, if the otolith is sedated with drugs during flight, the illness may be stopped altogether. A technique to prevent space adaptation syndrome would not only lead to more enjoyable flights for space travellers, it would also increase the effectiveness of many space missions, work on which is bound to suffer if the people inside the spacecraft spend a significant portion of their time feeling unwell.

In other research related to astronauts' health on the first Spacelab flight, scientists tried to find out why people often come back from space flights suffering from anaemia. It appears that something happens during weightlessness to interfere with the production of red blood cells. Workers in space medicine think they have pinned this down to a particular hormone that is present naturally in the blood and attacks cells. In normal circumstances, production of this hormone is triggered when the body realizes that, for any reason, it has too much blood and wants to get rid of some of the fluid.

Scientists involved in the Spacelab flight found that the blood streams of the astronauts contained twice as much of this hormone as usual. It is thought that this is because under zero gravity the upper reaches of the body contain more blood than is normal – on Earth, the blood is concentrated towards the lower parts as a result of gravity. The top elements of the body are, therefore, fooled into thinking they have too much blood, so initiating extra production of the hormone and a lowering in red-blood-cell levels. On future flights, it may be possible to give drugs to the astronauts to ensure that the hormone does not overproliferate.

An example of how medical experiments in space could benefit those of us who simply lead terrestrial lives came in a Soviet flight in 1984. Rakesh Sharma, the first Indian astronaut, who was present on this mission as a guest, did yoga exercises while special measuring instruments monitored his muscles. Results from experiments such as this could shed new light into some kinds of muscular diseases, or provide clues to the rehabilitation of people suffering from muscle disorders.

In botany, researchers in space workshops have the chance to study new conditions in which plants may grow more vigorously. Cosmonauts on the Soviet space station Salyut 7 showed that there was no problem in growing plants in space. They harvested at least a dozen different varieties, including peppergrass, wheat, onions, parsley, radishes, oats and peas.

Astronomers and remote-sensing specialists are among other types of researcher who can see the possibilities of space stations. Such facilities are, after all, just big satellites with provision made for people. With satellites, astronomers have for years gathered information about the universe by putting instruments on them and pointing them at the heavens. In a similar way, remote-sensing researchers have plied their trade with satellites. These have taken the form of travelling vantage points from which to collect information about events or conditions on Earth.

The concept that people may be able to go aloft with the instruments on manned space stations holds out new opportunities. Those concerned may be able to pick up a few snippets of information in addition to enjoying the ride. It would also add flexibility to the venture – the instruments could be manually adjusted if anything unforeseen happened.

A space station, for similar reasons of flexibility, appeals to the military. It acts as a very good position from which to monitor the activities of potential enemies or from which to wage war. What is more, with people in space, a military power will be capable of a far wider range of possible actions than with a space base used as a platform simply for hardware. (See Chapter 6 for more about this particular application of the space station.)

In the rather more distant future, men and women may travel away from the Earth not just to stake out what are, in effect, prime pieces of industrial, scientific or military real estate, but also to undertake space ventures to seek out something still more concrete – fresh supplies of raw materials. Some of the world's space planners point out that the Moon and other planetary bodies contain huge repositories of metals and other materials of which the Earth is running short. According to the more enthusiastic of these technological soothsayers, the Earth's near neighbours in space may, within a few decades, resonate to the sound of mechanical diggers and drilling equipment as mining companies exploit riches that are literally out of this world. Space tugs could transport the raw materials to the orbital factory complexes which would already be in position and from which would flow the novel products that end up in the shops and warehouses on Earth. (For more about the Moon factory, see Chapter 8.)

All this may, of course, be little more than dreaming, but from the earliest days of space exploration observers have speculated about the future, and the idea of space stations and new sources of metals and other materials has often loomed large in their thoughts. Keith Glennan, the first administrator (director) of NASA, told a Senate committee in February 1959, jut six months after his organization had been set up:

> We do not know the ultimate role of space vehicles in transportation, any more than the few spectators of the early flights of the Wright brothers knew of our present jet transports that make the world a neighbourhood. Some speculate that the Moon and the planets represent a vast new physical frontier, a source of material wealth.

In a book published in 1957, Eric Burgess, a former chairman of the British Interplanetary Society, stressed the scientific uses of a manned space station. Burgess may have been prosaic, dull almost, but he accurately predicted what scientists would be talking about a quarter of a century hence: 'Generally, the station will be used for astrophysical work and for studies of geophysics, free-fall biology and other physical studies needing human supervision and which cannot be undertaken in laboratories on the Earth's surface.'

Meanwhile, Arthur C. Clarke also talked of space stations in his seminal *The Exploration of Space*, published in 1951. He thought that the orbiting complexes would find their main applications in astronomy and Earth surveys. Perhaps surprisingly, the book barely mentions the notion of putting industry heavenward. But Clarke did have one unusual idea for space stations – as free-flying rest homes or hospitals:

> The fact that, apart from relatively short periods of eclipse by the Earth, a space station would be in continuous sunlight would be of great therapeutic value. So would be the spectacle of the Earth itself, almost filling the sky and going through its phases from new to full in a few hours. The infinite variety of detail presented by the continents, seas and clouds, the pleasure of picking out familiar landmarks and even of observing the streets of the great cities by telescopes, should reconcile the patients to their temporary exile. The 'Earthside' ward of a space-station hospital would, indeed, be a room with a view!

Such visionary thoughts were, we may assume, far from the minds of the engineers from the USA and USSR who, in their planning for the first manned flights of the 1960s, laid the ground for the purely practical work in space stations. These initial steps in putting people into space were taken very much in the spirit of adventure. The chief motivation was to explore regions into which no one had been; to demonstrate activities that, previously, had been thought impossible.

Only after the early manned flights did people start to think in concrete terms of the applications that could stem from trips out of the atmosphere. In much the same way, we can assume, the pioneering seafarers of 500 years ago made their initial voyages simply for the experience of doing something new, but in the process they opened up new sea routes and discovered new land masses. The preliminary explorations of these people were capitalized upon by later generations of merchants and industrialists.

The link between the beginning of the maritime era and the activities in twentieth-century space technology was captured nicely in the title of a project by researchers in West Germany and Italy that started in 1983. The study looked at the ways in which Western Europe and the USA could collaborate in building a permanent, manned space station. The project was called 'Columbus'. 'Europe discovered America in 1492,' said one official in the space department of West Germany's research ministry, 'and we think that 1992 is a suitable date for another mission of discovery, putting into orbit a space station.'

Western Europe and the USA have, however, a long way to go before they match the Soviet Union in this area. This country dominates the scene when it comes to keeping people in space for long periods in orbiting complexes. The first person in space was a Soviet citizen. Yuri Gagarin, aged twenty-seven, an officer in the Soviet Air Force, made one circuit of the Earth inside his 5-tonne Vostok 1 spacecraft on 12 April 1961. Vostok was put into orbit by an A-1 launch vehicle based on the SS-6 Sapwood ballistic missile. The complete trip lasted 1 hour 48 minutes. Gagarin stayed cool during his monumental achievement. 'The sky,' he recounted, as he flew around the globe, 'looks very, very dark and the Earth is bluish.' A little under one month later, the USA put its own astronaut into space. Alan Shepard made a fifteen-minute sub-orbital hop above the atmosphere on 15 May on another converted missile.

Soon after this, the engineers (and the politicians behind them) stepped up their efforts to examine exactly what they could achieve by blasting people into the cosmos. One goal was clear – to put a man on the Moon. As the history books show, the USA

Table 13. Summary of manned space flights (to 31 December 1984)

	USSR	*US*	*Total*
Manned missions	57[1]	45	102
Astronaut flights	177	131	248
People in space	69[2]	86[3]	155
Visits to space stations[4]	29	3	32
Moon landings	0	6	6
Total flight time (person days)	3,755	1,322	5,077

1 Soviet figure includes Soyuz 18A flight in 1975 in which mission aborted after 15 minutes, but *not* initial Soyuz T-10 mission in 1983 in which rocket exploded on launch pad.

2 Total of 58 Soviet citizens, plus 11 guest cosmonauts – one each from Czechoslovakia, Bulgaria, Romania, Hungary, East Germany, Poland, France, Vietnam, Mongolia, Cuba and India.

3 Total of 84 US citizens, plus two 'guests' from West Germany and Canada.

4 Figure includes occupations by crews, not dockings without occupation.

accomplished this with a total of six lunar landings between 1969 and 1972. (Details of the Moon landings, together with a summary of manned missions to date, are given in Tables 13 and 14 and Figure 7.)

The landings on the Moon could not have been achieved without a mastery of one key technology – the ability to dock different spacecraft in orbit. However, while the USA perfected this to achieve its aim, the Soviet Union, if anything, exploited the principle rather more fully. It did this not by venturing onto the lunar surface but by harnessing the idea of joining up spacecraft to

Table 14. Major landmarks in manned space flight

1961	Vostok 1 (USSR)	first manned Earth orbit (Yuri Gagarin)
1962	Mercury 7 (US)	first US Earth orbit (John Glenn)
1962	Vostok 3 and 4 (USSR)	first dual space mission (Andrian Nikolayev and Pavel Popovich)
1963	Vostok 6 (USSR)	first woman in orbit (Valentina Tereshkova)
1964	Voskhod 1 (USSR)	first three-man orbit mission (Vladimir Komarov, Konstantin Feokistov, Boris Yegorov)
1965	Voskhod 2 (USSR)	first space walk (Alexei Leonov)
1966	Gemini 8 (US)	first docking in space, between Gemini vehicle and unmanned target craft (Neil Armstrong and David Scott)
1967	Soyuz 1 (USSR)	first flight of Soviet second-generation space vehicle; crewman (Vladimir Komarov) died after technical failure
1968	Apollo 8 (US)	first manned orbit of Moon (Frank Borman, James Lovell and William Anders)
1969	Soyuz 4 and 5	first docking of two manned spacecraft (Vladimir Shatalov, Boris Volynov, Alexei Yeliseyev, Yevgeni Khrunov)
1969	Apollo 11 (US)	first lunar landing (Neil Armstrong, Eugene Aldrin)
1971	Soyuz 11 (USSR)	first docking with space station (Salyut 1); crew died on return trip (Georgi Dobrovolsky, Vladislav Volkov, Viktor Patsayev)

1971	Apollo 15 (US)	first astronauts to drive on Moon (David Scott, James Irwin)
1973	Skylab (US)	enters orbit
1975	Soyuz 19 (USSR) and Apollo 18 (US)	first (and only) docking between Soviet and American vehicles (Alexei Leonov, Valeri Kubasov, Thomas Stafford, Vance Brand, Donald Slayton)
1977	Salyut 6 (USSR)	enters orbit
1977	Soyuz 26 and 27 (USSR)	first double docking with space station (Salyut 6). (Yuri Romanenko, Georgi Grechko, Vladimir Dzhanibekov, Oleg Makarov)
1978	Progress 1 (USSR)	first trip of unmanned Progress freighter to space station (Salyut 6)
1980	Soyuz T-2 (USSR)	first flight of improved Soyuz craft (Yuri Malyshev, Vladimir Aksyonov)
1981	STS-1 (US)	first flight of space shuttle (John Young, Robert Crippen)
1982	Salyut 7 (USSR)	enters orbit
1982	STS-5 (US)	first deployment of satellites from shuttle (Vance Brand, Robert Overmyer, William Lenoir, Joseph Allen)
1983	STS-9 (US)	first launch of Spacelab and first six-man crew into space (John Young, Brewster Shaw, Robert Parker, Owen Garriot, Byron Lichtenberg, Ulf Merbold)
1984	STS-13 (US)	first seven-person crew into space (Bob Crippen, Jon McBride, Kathryn Sullivan, David Leestma, Sally Ride, Paul Scully-Power, Marc Garneau)

the construction of orbital stations that could act as prototypes for factories.

During the US lunar missions, three astronauts were transported close to the Moon's surface, where the mother craft separated into two. One vehicle stayed in lunar orbit, while the other landed with two men on board. On the return trip, the two capsules had to dock in orbit around the Moon so that all three astronauts could get back to Earth safely. The USA practised for its trips to the Moon with a series of 'dummy runs' near to the Earth. The first space docking was in March 1966. Neil Armstrong and David Scott, the astronauts

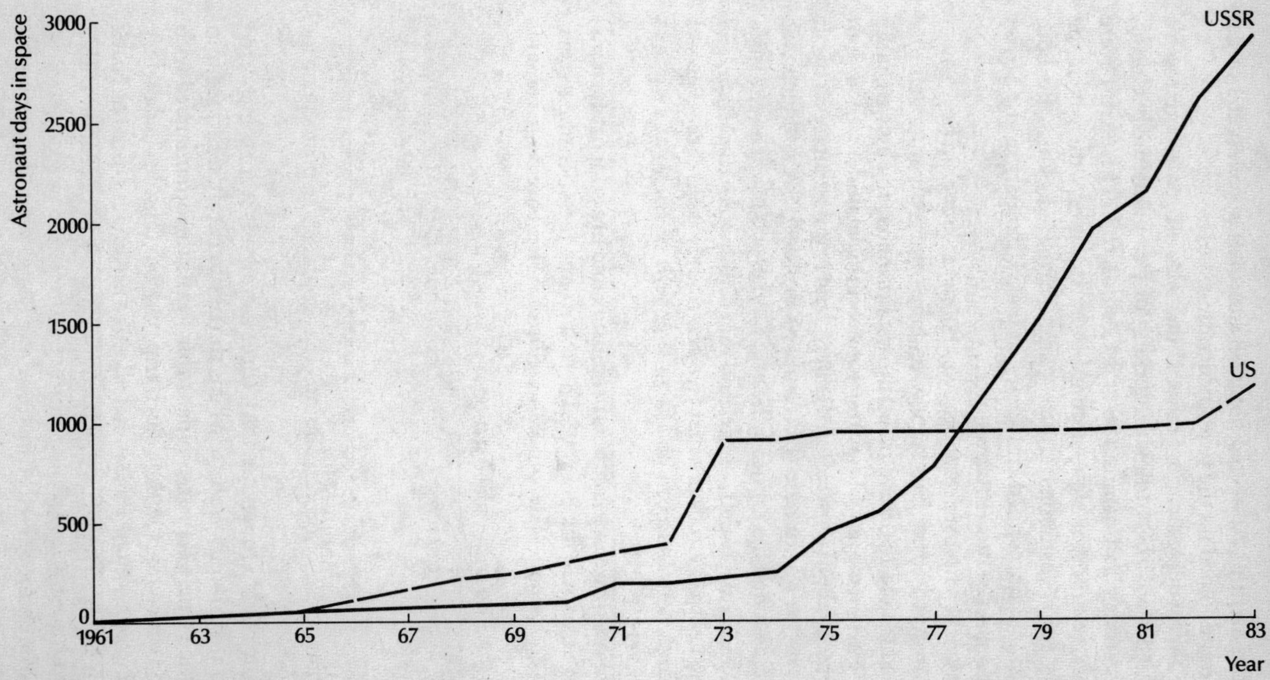

Figure 7. Cumulative astronaut days in space 1961–83

on board Gemini 8, steered alongside an unmanned Agena target vehicle and joined up the two spacecraft.

The following year, the Soviet Union also scored an impressive 'first'. This was the automatic docking of two unmanned vehicles, known simply as Cosmos 186 and Cosmos 188. These craft were pilotless versions of the 7-tonne Soyuz capsules with which the Soviets ferry cosmonauts into space. Further dockings by both manned and unmanned craft were to play a big part in the Soviet Union's space station programmes.

The world's first space station was Salyut 1, launched by a Soviet rocket in April 1971. This station, and the other Salyut models that followed (Table 15), have basically the same shape. They are large metal cans made up of several interlocking cylinders. A Salyut spacebase is about 4 metres in diameter and some 15 metres long. It weighs 15 to 20 tonnes. (By comparison, the USA's Skylab was much bigger and four times as heavy. Skylab was based on the third stage of a Saturn rocket, of the kind that sent men to the Moon, and launched with the rocket's first two stages. It housed three crews of astronauts between 1973 and 1974 for spells of 28, 59 and 84 days respectively.)

A Salyut space base contains propulsion and life-support systems, plus solar panels for energy. There is room for scientific experiments and small-scale process facilities. The base also contains

Table 15. The Soviet space stations

	Launch	*Destruction*[1]	*Occupations by cosmonaut crews*
Salyut 1	April 1971	Oct. 1971	1
Salyut 2	April 1973	May 1973	—
Cosmos 557[2]	May 1973	May 1973	—
Salyut 3	June 1974	Jan. 1975	1
Salyut 4	Dec. 1974	Feb. 1977	2
Salyut 5	June 1976	Aug. 1977	2
Salyut 6	Sept. 1977	July 1982	14
Salyut 7	April 1982	—	7[3]

1 By loss of orbit and burning up by friction in the atmosphere.
2 Failed Soviet space station.
3 To end of 1984.

accommodation for a crew basically of two, though six can be squeezed in. A vital part of the stations are their docking sections, to which other spacecraft sent up from Earth can be linked.

The first space ship to attempt a docking with Salyut 1 was Soyuz 10, which left the Earth a few days after the station had gone into orbit some 300 kilometres above the planet. The three cosmonauts, Vladimir Shatalov, Alexei Yeliseyev and Nikolai Rukavishnikov, steered their capsule close up to the target but failed to make a rendezvous. Observers speculated afterwards that either one of the crewmen had become ill or that the mechanical systems which were meant to connect the two spacecraft had failed. In any event, the trio of spacemen returned to Earth, their mission unaccomplished. They followed the conventional pattern of Soviet space flights by landing inside a return capsule (a small part of the Soyuz craft in which they had originally entered space), whose descent into the central part of the Soviet Union was braked by parachutes.

The next crew dispatched by the Soviet Union had better fortune, at least initially. A couple of months after the previous team had come home, Georgi Dobrovolsky, Vladislav Volkov and Viktor Patsayev left the USSR's Tyuratum space base aboard Soyuz 11. This time, the docking procedures passed off successfully. On 6 June 1971, the cosmonauts became the first men to inhabit a 'space hotel'. They stayed on board for three weeks, studying the stars and the Earth's surface with optical telescopes and doing experiments in botany and genetics. All went well until shortly after the crew clambered back into their Soyuz spacecraft, which had remained all the time docked to the larger Salyut vehicle, to begin the journey back to Earth.

On the point of re-entry into the planet's atmosphere, disaster occurred. An explosion forced open an air lock in the return capsule. The air from the pressurized cabin spurted out and the crew, who were not wearing space suits, quickly expired. Salyut 1 was never reoccupied. In October 1971, it began to run out of the propellant needed to keep it in orbit. Shortly afterwards, the space station lost altitude, entered the atmosphere and burned up as a result of friction.

The Soviets took almost three years to recover from their ill-fated efforts with the first Salyut. Unfortunately for the USSR, Salyut 2 failed soon after launch. It was not until a two-man crew entered a new station, Salyut 3, in June 1974 that the USSR's space engineers recovered their confidence. The latest Salyut was an improved version of the first. It was about the same size but had better living quarters, and a more sophisticated system for collecting energy from the Sun with solar panels.

Salyut 3 was probably mainly for military applications, for example surveillance of the Earth with high-resolution cameras. The two crews sent to the station (only one actually got inside; the other failed to dock) consisted of military officers, and when communicating with mission controllers on the Earth, the cosmonauts used code to reduce the risk of eavesdroppers.

Engineers in the USSR practised further manoeuvres with two more space stations, each of which saw visits by two crews of spacemen. The longest of these, in 1975, lasted for a little more than two months, but the Soviets kept their most impressive feats for the period after September 1977, the date that marked the twentieth anniversary of the launch of the first Sputnik.

To commemorate this event, a Soviet booster launched a new station, Salyut 6. This featured several improvements over its predecessors. Most importantly, it had two docking ports, compared with one on the previous stations. As a result, the Salyut could play host to an unmanned supply ship called a Progress freighter, at the same time as remaining attached to a Soyuz craft. The Progress vehicles carried supplies from Earth which helped to eke out long-stay missions.

Alternatively, as a result of the twin ports on the new space base, two Soyuz craft could dock simultaneously. This led to the general pattern in which spacefarers on long missions of several months acted as 'innkeepers' in space, hosting short visits of a week or so by 'guest' cosmonauts who would dock in the empty port.

During the four-year life of Salyut 6, the Soviets several times broke the record for long-endurance space missions. Crews stayed inside the space base for 96, 140, 175 and then 185 days. Salyut 6 was visited by no fewer than sixteen Soyuz vehicles, giving a total of twenty-seven people with experience of life in space. Five of the crews stayed for two months or more; the others generally remained for about a week. The short-stay missions were of doubtful technical or scientific value. Their main function was to take into space non-Soviet cosmonauts from other communist nations in what were little more than publicity exercises. In this way, representatives from Vietnam, Cuba and Mongolia as well as several East European countries came to regard Salyut 6 as their home for short periods. Probably of greater long-term significance were the twelve visits to the station of the unmanned Progress craft, which ferried into space some 22 tonnes of freight and equipment.

Another new development was the application by the Soviets of a new class of manned space ships, called Soyuz T ('T' stands for transport). These could carry three people quite comfortably. The ordinary Soyuz vehicles which they gradually replaced were meant only for two, though three could be accommodated in a tight fit. The new 'T'-type craft also had improved propulsion, an automatic docking system and their own solar panels to increase endurance.

A further crucial event came after the last crew to inhabit Salyut 6 had returned home. Leonid Popov and Dumitru Prunariu (a 'guest' cosmonaut from Romania) left the station in May 1981. The next month, an unmanned ship called Cosmos 1267 docked in one of the station's ports. Mysteriously, the vessel had orbited the Earth for a couple of months before approaching the space station. Cosmos 1267 had an initial take-off weight of some 15 tonnes, twice that of an ordinary Soyuz vehicle. According to space analysts in the USA, before docking with the Salyut the vessel ejected to Earth a re-entry vehicle that could be a prototype of some kind of Soviet 'mini shuttle', a winged space plane that, like the USA's shuttle, can return on a runway rather than land by parachute. After a series of tests of such mini shuttles, the Soviet Union admitted in December 1984 that such vehicles could be a useful way of returning people or materials to Earth from future space stations.

The combined Salyut 6–Cosmos 1267 complex had a mass of some 30 tonnes. Almost certainly, said observers, this was a step by the Soviets towards linking up a number of craft in space to give bases still larger than the orthodox arrangement of a Salyut station with two Soyuz transporters at either end. The docked complex continued to circle the Earth until July 1982, when ground controllers instructed it to re-enter, causing the hardware to break up. But by then the Soviets had already launched a new station which took up position in April 1982. Salyut 7 continued the step-by-step improvements, but it was very similar to the previous space base, and the pattern of visiting and long-stay crews continued. In 1982, one of the teams of cosmonauts, Anatoli Berezovoi and Valentin Lebedev, set up a new endurance record by staying aboard the station for 211 days.

In May 1983, the Soviets played yet another important card. A large vessel called Cosmos 1443 was docked with Salyut 7. This weighed some 20 tonnes, similar to the space station itself. Salyut 7 already had a Soyuz transporter attached to the other port, making the total mass of the space complex about 50 tonnes.

Cosmos 1443 was, in reality, two capsules joined together. In total it was 14 metres long and 4 metres in diameter. The first capsule was, in effect, a long room that could be added to the Salyut's own facilities. It had its own control and life-support system for the crew, plus 40 square metres of solar panels. This part of Cosmos 1443 also had a propulsion unit so that, when docked to the Salyut, it could shift the whole complex to different orbits. The second part of the new spacecraft was still more unusual. It was a descent module, similar in principle to the return capsules on Soyuz vehicles. With this, cosmonauts could send back to Earth photographs and documents. This segment of the Cosmos vehicle filled a gap in the plans by the Soviet Union for vehicles that could service complexes in space. Although the unmanned Progress vehicles can take items up to space stations, no Progress has ever made it back to Earth because it lacks the re-entry capability.

All the signs are that the docking was another practice for constructing stations by joining up different modules in the style of Lego building blocks. Cosmos 1443 stayed linked to the station until August 1983. After this, although there was no official confirmation, part of it probably re-entered the atmosphere and was destroyed. The re-entry segment of Cosmos 1443 returned to Earth with a cargo of documents and research material. It looks likely that the USSR will put into orbit further versions of the Cosmos vehicle and so consolidate its capabilities in orbit.

During the early part of the life of Salyut 7, the Soviets showed, too, that they were interested in keeping their stations permanently occupied. The other Salyut bases had always spent at least part of their time uninhabited; departing crews had to shut down or 'mothball' the station, for example switching off electrical equipment and de-activating experiments. A new team of cosmonauts would then visit the empty station a few weeks, or months, later and restart everything.

In September 1983, however, Vladimir Titov and Gennadi Strekalov were poised to become the first space-station relief crew. They were due to take over from Salyut 7's current occupiers, Vladimir Lyakhov and Alexander Aleksandrov, who had been aboard since the previous June. This strategy would have kept the station ticking over continuously, but in fact the plan failed to come off. The rocket carrying Titov and Strekalov exploded on the launch pad, forcing them to abort and parachute to safety; they were within seconds of losing their lives. Titov and Strekalov must have felt they were fated. A few months before, they had made an earlier attempt to reach the (then unmanned) Salyut 7, but had been forced to return to Earth in their Soyuz capsule after a docking mechanism failed, an event that is an occupational hazard in the business of visiting space stations.

After a 149-day stay in space, Lyakhov and Aleksandrov eventually returned to Earth in November 1983. This followed a minor scare occasioned by the ageing nature of their Soyuz ferry craft, which had stayed attached to the Salyut for longer than normal. Generally, cosmonauts come back to the USSR in a Soyuz brought to the station by a visiting crew. In this way, the vehicles stay in space no longer than is absolutely necessary. During prolonged spells in space, fuel can leak out from the vehicles and their fabric may also be damaged by the rapid changes of temperature caused by the space complex entering successive periods of sunlight and shadow.

Whatever the problems over getting a relief crew into orbit, there was every sign that the Soviet Union would soon try again the idea of keeping a station permanently occupied. In February 1984 three more Soviet cosmonauts, Leonid Kizim, Vladimir Solovyov and Oleg Atkov, embarked on a long-term occupation of Salyut 7. During their stay they carried out an intensive programme of scientific research. It was reported that attempts were being made to produce biologically pure substances for use in medicines. Several other successes were achieved, notably the link-up of five unmanned supply vehicles, the visit of two manned craft to the station and six spacewalks by two of the crew in order to repair the propulsion system. In July, Svetlana Savitskaya, while visiting the craft, became the first woman to make a second mission into space and also the first woman to make a spacewalk. Kizim, Solovyov and Atkov returned to earth in October 1984 after a total of 237 days in space. This was twenty-six days longer than any previous mission (Table 16). According to Soviet reports the cosmonauts had coped well with their long stay in zero gravity.

News has slowly filtered out of Moscow concerning the Soviets' long-term plans for space stations. A good pointer to the future was given in October 1982 in a statement by Vladimir Shatalov, a former cosmonaut who is in charge of the training of the USSR's space men and women. Shatalov spoke of the philosophy of assembling stations in orbits in modules. It was difficult to foresee, he said, how rockets could put very large complexes into space in one launch. A station with a mass of 1,000 tonnes would require a rocket more than 500 metres tall with a launch weight of tens of thousands of tonnes. 'Clearly,' said Shatalov, 'this way is hardly acceptable.' He went on:

> A more expedient approach is to assemble in orbit a large complex out of individual blocks, each of which can operate independently of others. A block is orbited, and the station starts to function. A while later it is joined by another block, still another, and so on. In that way it would be possible to build up a space station of any size and any purpose – research institutes, or factories to furnish the Earthmen with unique materials and medicines.

Invoking the spirit of Konstantin Tsiolkovsky, the Soviet space theoretician mentioned at the beginning of Chapter 2, Shatalov said that one day it might be possible to build from these modules 'ethereal settlements', with room for hundreds of people. Such

Table 16. Astronaut records

Most trips into space	John Young (US) with six flights; Gemini 3 (1965), Gemini 10 (1966); Apollo 10 (command module pilot, lunar mission, 1969), Apollo 16 (lunar landing, 1972); two space shuttle flights (STS-1, 1981; STS-9, 1983)
Longest spell in space	237 days by Leonid Kizim, Vladimir Solovyov and Oleg Atkov (USSR) (February–October 1984)

settlements in the heavens would operate in a closed ecological cycle: they would grow vegetables, rear animals and regenerate air and water.

According to one noted watcher of Soviet space activities, the Dutch journalist Peter Smolders, the next USSR space station will probably be launched in 1985 or 1986. The station's basic core will be roughly the same as Salyut 6 or Salyut 7, but it will have a total of six docking ports as opposed to the two on these earlier stations. Two of the ports would be reserved for Soyuz and Progress craft, following the established pattern. To the other four would be attached a series of new modules for specific purposes. The modules would be for observations in astronomy; ventures in low-gravity materials processing, for example electrophoresis or crystal growth; surveys of the Earth with high-accuracy cameras; and experiments in biology. Smolders has produced a diagram (Figure 8) of what he says this station will look like. He produced this after conversations with members of the Soviet design team.

The lead that the USSR has in space stations has worried the USA. After America won the race to reach the Moon in the 1960s, its space programme entered a lean period. In the 1970s, the country's space activities were confined to a few more Moon missions as the Apollo programme tailed off, the Skylab flights, plus one joint venture in space with the Soviet Union in 1975 in which craft from the two nations linked up in orbit.

The USA's spell in the doldrums may, however, have come to an end, thanks mainly to the $15,000 million development of the space shuttle. This is essentially a re-usable 'truck' for lifting payloads into low orbits some 300 km above the Earth. Most of these payloads will, in the early years of the shuttle's operation, be satellites. (Some of them need a small rocket, an 'upper stage', with which they are boosted from the low altitude to which the shuttle can fly to much higher orbits such as the geostationary ring 36,000 km high.) There will be not just one shuttle but several. NASA plans to have four in operation by late 1985, of which three – Columbia, Challenger and Discovery – had entered service by 1984.

It has dawned on the USA's space planners that the shuttle would also be a good tool for building space stations. In much the style of the Soviet Union, the vehicle would take up modules that would later be docked to form large habitats. Whereas the USSR can lift

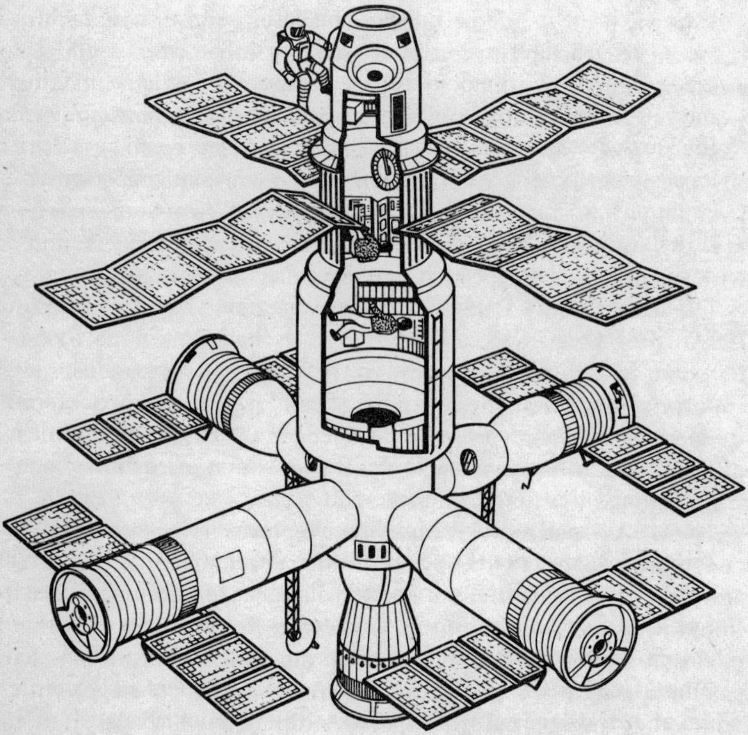

Figure 8. Plan of a future Soviet space station

loads of only up to 20 tonnes with its expendable rockets (though more powerful launchers of higher capacity may be on the Soviet engineers' drawing boards), the space shuttle can take one-and-a-half times as much up to a low orbit. Furthermore, a large part of the shuttle is re-usable and so, ultimately at least, should be cheaper.

Officials at NASA are considering several options for the US space station, which would have applications not just in science and industry but would also do jobs for the Department of Defense. At the end of 1983, NASA finished the design of the station which it would like to see enter orbit around 1991. The base would follow along the lines of the Soviet space stations. It would be built from modular sections fitted together in orbit and carried separately from the Earth. The advantage that the USA has over the Soviet Union is its possession of the space shuttle, which is an ideal vehicle for lifting components into low orbits. According to Bob Freitag, the deputy director of a special space station 'task force' within NASA, the base's modules would each have specific purposes, for example for different types of scientific experiments or as vantage points from which to repair or service satellites.

The first phase of the station, which would house six to eight astronauts, would cost some $8 billion. Later, the station would 'grow', with other units being added. One of the basic building blocks for the station could be versions of the Spacelab modules developed by the eleven nations of the European Space Agency. The laboratory, an aluminium cylinder with room for several people, fits into the cargo bay of the space shuttle, and this circled the Earth for ten days on its first mission in 1983, a joint exercise between NASA and ESA. Officials from both agencies were delighted with the results from the venture. Of seventy-one experiments – some housed inside the pressurized cylinder and others packed into a pallet outside that can be exposed to the vacuum of space – all but one was a success. The experiments on Spacelab's first mission included ventures in materials processing, biology, astronomy and the study of human behaviour in weightlessness. For the first time, the shuttle which carried Spacelab into orbit had a crew of six people. Besides four specialist astronauts (these included the veteran John Young, who was making his sixth trip into space (Table 16)), the crew included two professional scientists, Ulf Merbold from West Germany and Byron Lichtenberg from the USA. These people tended to do most of the experiments inside the pressurized module, which was connected to the crew quarters of the shuttle by a 'crawl-through' tunnel (Figure 9). On future Spacelab flights, such scientists, known in NASA jargon as 'payload specialists', will play a large part. They will be plucked from Earth-bound laboratories, given only minimal training for space and go aloft for a week to ten days to supervise scientific ventures.

At present, Spacelab cannot exist on its own in space. For life-support systems, power and propulsion, it relies on the shuttle and so must spend all its time in orbit inside the cargo bay of the American launch vehicle, but future generations of the module could exist independently. According to planners in Western Europe, they could form some of the modules that make up the US space station of the 1990s. Another promising notion is to use in the station another element of hardware that Europe is designing. This is a small unmanned platform called Eureca which should make its first flight in 1988. Like Spacelab, this will be taken into space by a shuttle, but there the launch vehicle will abandon the platform, leaving it to exist in orbit for up to six months before a shuttle returns to collect the hardware. Eureca will carry about 1 tonne of experiments, supervised either by automatic equipment or by commands from controllers on Earth. In a space station of the 1990s, Eureca could form the basis of small space tugs that fly between elements of the base, or perhaps carry loads to the main core of the station from shuttle supply vehicles.

Space planners in Western Europe are anxious to capitalize on the work on Spacelab by becoming involved with NASA's plans for the space station. In any future joint venture with the USA, Western Europe will be more of an equal partner. This contrasts with the arrangement with Spacelab, where ESA was very much the junior member of the team. The Spacelab story goes back to 1973, when Western Europe had very little experience in space technology. European governments and the USA hatched a deal. In exchange for one free ride on the shuttle (then, of course, still unbuilt), Western Europe would build a space laboratory that could be used time and again. Under the terms of the deal, the first Spacelab mission would be a joint undertaking between the USA and Europe, with each side contributing roughly equal amounts of experiments, but afterwards the hardware was to remain the property of the Americans, who

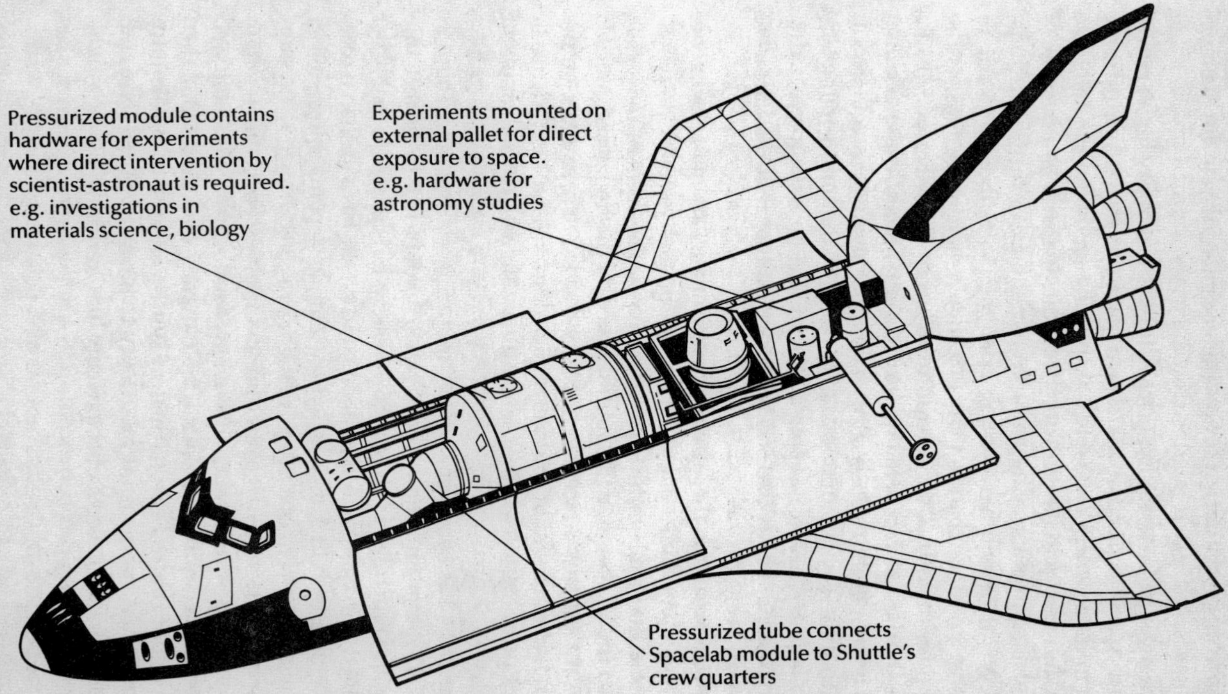

Figure 9. Spacelab

would charge Western Europe the commercial rate for using it again.

The arrangement was the price that ESA had to pay for its novice status. Spacelab has cost the eleven countries in the agency some $750 million, with West Germany putting up more than half. The USA has also purchased from European industry another set of Spacelab equipment, at a cost of $200 million. For future missions, however, ESA will have to pay around $180 million for a trip in space of no more than ten days. Roughly half of this cost is the launch fee that NASA will require for hiring the full cargo bay of a space shuttle, the rest being the cost of experiments, training and so on. The first flight of Spacelab cost a total of about $330 million, not counting the development costs of the hardware but including the full bill to NASA for keeping a shuttle in orbit for ten days. This worked out at about $20,000 for every minute that the hardware stayed in orbit.

Not surprisingly, these huge sums of money are making government planners think hard about hiring Spacelab for more missions. West Germany is committed to its own Spacelab flight in 1985, code named the D-1 mission. Though shocked by the high price, officials in West Germany's research ministry decided to soldier on, on the grounds that the flight could give the country useful experience in an area that may lead to promising new industries in a couple of decades.

The USA has firm plans for two more Spacelab missions, both in 1985. If it can find the money, NASA would like to finance one Spacelab launch a year after 1986. Western Europe will probably get involved with more Spacelab flights, but on the basis of shared costs with either the USA or Japan. For example, Japan is interested in funding 40 per cent of a Spacelab flight in 1988.

Experiments in space do not come cheap. Western countries, hard hit by economic recession and a range of social problems, could well decide that they should spend their money elsewhere. But then they would run the risk of missing out on an area of opportunity that promises to open up exciting new horizons in industry and technology in the world of the twenty-first century.

8 ☆ Enter the Robot

No matter how perfectly instrumented a satellite may be, it cannot be perfect unless there is a thinking being inside, capable of observing, weighing evidence, making decisions and carrying them out.

Scott Carpenter, former American astronaut

In all the history of space projects, you have rarely needed people. If you look at the Moon landings, the scientific knowledge obtained by automated devices was orders of magnitude greater than that obtained by debriefing the astronauts.

Berndt Feuerbacher, researcher at the European Space Agency

Which of these two views is correct? For twenty-five years, the world's space planners have argued about whether people or machines should shoulder the burden in exploiting the riches of the Universe. In fact, both views will be reflected in the space programmes of the coming decades. Men and women will not be eliminated from space operations, but much of the hard work will be done by robots and other automated devices which have some degree of ability to take decisions for themselves. Humans will enter the picture by playing a key supervisory role. The machines will generate their own activities to a limited degree, leaving people to intervene, when necessary, from control points perhaps thousands of kilometres away. In this way, humans will remain very much in charge of space projects, but they will generally dictate the course of events not from space itself but from the comfort and security of bases on Earth.

The debate over who or what should play the leading role in space projects has featured strong opinions. Some say that men and women must always go where the action is thickest in space, where the jobs are too important to be left to automatic circuitry. According to this view, astronauts and cosmonauts have often rescued missions from disaster by their ability to respond to unexpected events.

In 1970, inspired engineering on the part of the crew of Apollo 13 brought back the mission from the brink of calamity. After an explosion disabled the Apollo ship's command module, the three astronauts were left without a means of propulsion. They converted the craft's lunar module, which was intended to operate only near to the Moon's surface, into a 'lifeboat' that ferried the crew back to Earth.

In terms of the overall effect on a space project, the work by the first team of spacefarers to visit Skylab, the USA's space station, was still more important. Had it not been for the imagination of Pete Conrad, Joseph Kerwin and Paul Weitz in repairing the vehicle after it had suffered damage during launch, the valuable space station would have had no more use than a large chunk of metal drifting aimlessly in the heavens.

Nearer the present day, two Soviet cosmonauts, Vladimir Lyakhov and Alexander Aleksandrov, did a similar job on Salyut 7 in the closing weeks of their 149-day stay in 1983. According to reports in the West, the station had suffered a fuel leak. If just machines, rather than men, had been on board, the Soviets might have had to abandon the Salyut two or three years before the end of its intended life.

John Glenn, the first American to orbit the Earth, summed up the danger of relying on machines in a report to NASA after his 1962 mission:

> Even where automatic systems are still necessary, mission reliability is tremendously increased by having the man as a back-up . . . This mission [Glenn's own] would almost certainly not have completed its three orbits, and might not have come back at all, if a man had not been on board.

General Thomas White, the chief of staff of the US Air Force, in 1960 gave the military rationale for keeping people in the driving seat in space operations:

> Despite all the efforts of industry and the genius of science, certain limits are automatically imposed where the machine is involved. A machine cannot think, nor can it reason . . . in the event of military operations in space, man's presence and ability to perform in space could spell the difference between defeat and victory.

In spite of these sentiments, good economic, psychological and technical reasons exist for reducing the role of humans in day-today activities in space. As space systems become more sophisticated, more jobs will have to be done in the harsh environment immediately above the atmosphere and beyond. To date, most of the people who have ventured away from the Earth have been explorers first and workers second, but this emphasis will change as space stations become established and a whole array of routine space-based tasks present themselves. These will include the maintenance of satellites and the operation of processing plants on permanent space stations.

For people to do these jobs would be incredibly expensive. With today's technology, astronauts in pressure suits can work outside their space capsules (in what NASA calls extra-vehicular activity or EVA) for only a few hours of the day. As people become more practised in this activity, the period may rise to eight or ten hours by the 1990s. According to an estimate quoted in a NASA study in 1980, one hour's EVA costs some $10,000 per astronaut. To keep a person inside his spacecraft for the same time (including the cost of putting him in orbit and returning him safely) costs around $200, according to the same study. (These figures should be multiplied by at least a factor of two to put them in the prices of the mid-1980s.) If intelligent machines were to take over some of the jobs of humans in space, NASA would save between $500 million and $5,000 million a year by the turn of the century, according to research in 1978 at the Jet Propulsion Laboratory in Pasadena, California.

It is easy to see why humans are such expensive entities to maintain in space. They require a lot of cosseting. If the body temperature varies by more than 10 °C either side of the normal of 31 °C, the organs cease to operate and death occurs. At temperatures below 10 °C, limbs become stiff and work is difficult. Above 30 °C, a person finds mental or physical activity strenuous. Machines do not suffer from the same kind of restraints. Not only must manned craft contain sophisticated mechanisms for regulating temperature, they must provide their crews with other necessities. Each day, the human body requires about 1 kg of oxygen, 2.5 kg of drinking water, 0.6 kg of food and some 3 kg of water for personal hygiene. There must also be some mechanism for disposing of the body's waste products: a daily quantity of 1 kg of carbon dioxide, 1.5 kg of urine, 0.15 kg of feces and 2.5 kg of water. Humans must also be protected from the cosmic radiation present in space, particularly in high orbits, that will harm their bodily mechanisms.

The extra hardware needed to cope with the fragile fabric of the human body puts up the cost of space missions. Furthermore, when people are involved in space missions, the risk of failure must be reduced to minute levels. If a human dies in space, or on a journey to or from it, the event is a national disaster. In political terms, such occasions can reduce public support for space programmes and put back missions above the atmosphere by several years. Both the Soviet and American space programmes have suffered severe setbacks as a result of deaths of astronauts. In 1967 and 1971, in two incidents, a total of four USSR cosmonauts died when re-entering the atmosphere. In 1967, three US spacemen expired in a launch-pad fire. The special nature of manned flights pushes space planners to spare no effort in ensuring that the hardware works, in extra tests and in arrays of back-up systems for example. Such efforts again increase the expense of manned flights.

A spacefarer's job is not particularly safe or healthy. A study in the Soviet Union said that cosmonauts face about the same level of occupational risk as test pilots or professional boxers. Most of the people sent above the atmosphere by the Soviet Union and the USA have suffered no long-term damage to their well-being. The current thinking in the USSR is that, with proper exercise and medical attention, cosmonauts should not be harmed by spells in space of up to eight months. For longer periods, however, the exposure to radiation and the effects of weightlessness may take their toll. Among the consequences of living in space are: gradual loss of brain cells; a disruption to the blood-circulation system which causes, among other things, swelling of the legs; a change in function of the kidneys, causing weight loss; muscular atrophy due to elimination of weight load; and weakening of the bones as a result of the loss of phosphorus and calcium. The biggest problem for spacefarers, particularly those on short missions, is a condition similar to travel sickness known in the USA as space adaptation sickness. This leads to nausea and a general feeling of being unwell. It affects many cosmonauts and astronauts in the first few days of flight and is probably due to the body finding it difficult to adapt to weightlessness.

As if these general physiological difficulties were not enough to reduce sharply the queue of people volunteering for space flights, boredom can also be a problem. Some of the Soviet cosmonauts who have stayed several months on space stations have suffered psychologically. Extracts from the diary of Valentin Lebedev, who was in space for 211 days in 1982 aboard Salyut 7, show that life for space travellers can be distinctly unglamorous. Lebedev's account gave a picture of sleepless nights, a yearning for home and frequent brushes with other members of the crew or with supervisors on Earth. 'Against the background of mounting weariness,' Lebedev wrote, 'relations can slip badly. In working with ground control and even among the crew there are some sharp moments but we cannot lose our tempers. Otherwise, a split would widen.'

Although manned flights have become commonplace, not everyone would make a good astronaut or cosmonaut. The ground rules for this occupation were written in 1960 by Brigadier General Don Flickinger, the head of the US Air Force's office of scientific research. The stipulations may have been relaxed over the years, but the same general criteria still apply. Flickinger wrote:

> We must screen our personnel to find those who are experienced observers, who have demonstrated qualities of coolheadedness and resourcefulness in tight situations, who have good intelligence, who can tolerate various extremes of physical punishment, and who are stable, calm and confident. We must reject those who, although able to give a good account of themselves, do so primarily to prove something to themselves or the world.

Flickinger went on:

> Individuals with neurotic or psychotic tendencies should be most assiduously screened out and eliminated from consideration . . . Freedom from impulsivity is another desirable trait, for the spaceman must act economically when action is required yet refrain from purposeless activity when inaction is appropriate.

In addition to the biological and psychological phenomena which prevent men and women from being ideal for space missions, advances in technology are producing machines that have a level of 'intelligence', or decision-making ability, similar to humans. It no longer seems absurd to talk of robots that do people's work for them in space.

The technology of producing robots for space operations would draw heavily on programmes in advanced computers for purely terrestrial applications. In electronics laboratories around the world, researchers are striving to perfect computers that have some of the reasoning and decision-forming abilities of men and women. Over the next ten years, this hardware will be seen increasingly in offices, warehouses and factories. Machines of this kind, instead of following set patterns of activity written into programs, sense changes around them and, on the basis of these observations, make judgements. The discipline of researching into and producing such hardware is known as artificial intelligence. If a computer that operates according to artificial intelligence were made to supervise a mechanism such as a mechanical arm or even a free-flying spacecraft, then this entity could exist independently of commands or instructions from humans.

A study group set up by NASA to look at the applications of artificial intelligence put the case strongly for machines taking over some of the key tasks in space. In its final report, in 1980, the group said:

> We expect that machine intelligence and robotics technology will be a vital contributor to the cost-effective implementation and operation of the required systems [in future space missions]. In some areas, it will make the system feasible not only for technological reasons but also in terms of commercial acceptability and affordability.

In future, people and machines will have to work together. In the space operations of the decades to come, we shall see 'hybrid' arrangements in which arrays of intelligent or semi-intelligent machines go about their jobs. Some of the time they will work autonomously; on other occasions they will be controlled by human supervisors far from where the work takes place.

Robert Frosch, a former administrator of NASA who is now in charge of General Motors' research laboratories in Warren, Michigan, argued in favour of hybrids in an article in September 1983 in *Spectrum*, the journal of the USA's Institute of Electrical and Electronics Engineers. Frosch wrote:

> There is a kind of schism, a schizophrenia, in the space community about people and machines. There has been a split historically between those who insist on doing everything in space with people and those who insist on doing everything with machines. But there has been remarkably very little work done on how people can work with machines in space . . .

Frosch went on:

It seems perplexing to me that whenever planners discuss building structures in space, they view astronauts as common labourers . . . But it's not going to happen that way. It's too expensive to use astronauts as common labourers. We can't even afford that here on Earth . . . The other pole is to do without people at all – that is to wait until we have a completely automated plan for the exploration and exploitation of the Solar System. That is not going to happen either.

It is not necessary to assume that the machines one uses are either perfectly intelligent or completely unintelligent, and it is not that people are necessarily more reliable or less reliable than machines. However, the human–machine combination is more reliable than either of them alone. We have, in fact, the hybrid possibility of a few intelligent people working with a very large number of moderately clever machines.

The thinking behind man–machine 'hybridization' of space activities has been mapped out in great detail by researchers at the Massachusetts Institute of Technology. Under contract to NASA's Marshall Space Flight Center in Huntsville, Alabama, the researchers completed in 1983 a report on specific applications of machine intelligence in space missions. The MIT team, from the institute's Space Systems Laboratory and Artificial Intelligence Laboratory, sketched out the concept of 'telepresence' in which machines in space can operate with some degree of ability to make judgements for themselves but under the overall control of humans.

An example of a spacecraft with 'telepresence' could be a grappling device which would be able to repair satellites in orbit (Figure 10). The machine would be given rough instructions about what to do and be in continuous contact by radio links; but the equipment itself would have enough 'sense' to cope with unexpected situations. If, for example, a wrench in the device's robot 'hand' became stuck, the hand would automatically take the appropriate action (for instance by pulling hard at the wrench) without the controllers giving it specific instructions.

In another illustration of telepresence, automated hardware could be sent into space on missions of exploration. Such apparatus would be several steps more advanced than the robots that are commonplace objects in terrestrial factories. First, the equipment would have small rocket thrusters or would be on wheels so that it could move around. In this way, an unmanned space truck could zoom about between asteroids or trundle over the surface of Mars. Secondly, and still more importantly, the machinery would be able to react to

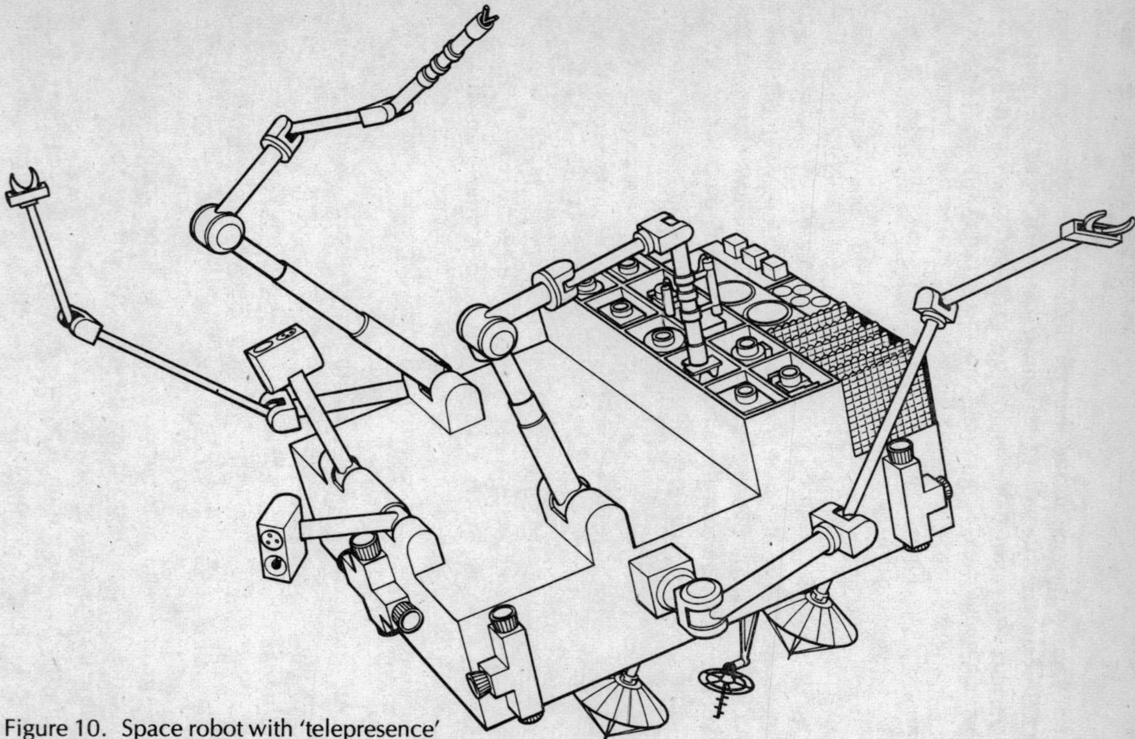

Figure 10. Space robot with 'telepresence'

unexpected events. Conventional industrial robots cannot alter their routine in the light of changing circumstances. If, for instance, the part that the machine wants to pick up moves into an unexpected position, the robot cannot adapt its operation to take this fact into account.

A space robot, on the other hand, would pick up information from its surroundings with sensors such as TV cameras or instruments that detect the texture of objects. The information so obtained would be sent to a central computer 'brain'; this works according to artificial intelligence and makes logical deductions to tell other parts of the robot what to do. In this way, an advanced computerized vehicle moving about the surface of a planet would automatically avoid obstacles such as boulders. It might even head off towards specific sites because it 'decides', on the basis of the available evidence, that they appear interesting (Figure 11).

A key aspect, however, is that the space machine would not be totally autonomous. A person would always be able to intervene, to lend that extra bit of flexibility and imagination that a piece of equipment, no matter how cleverly it is contructed, cannot be expected to possess. Built into the telepresence unit would be sophisticated feedback arrangements so that the operator, even though he may be far away, has a feeling that he is actually present.

The operator would not want, or need, to intervene more than intermittently. In the case of the robot explorer, humans would control the machine rather as a shepherd orders his dog to round up flocks of sheep. The shepherd would think it beneath his dignity (as well as a waste of time) to tell his charge the nuances about persuading woolly animals to run in different directions. In any case, the dog does not need telling. In a similar way, the space operator would give his automated machine an instruction such as 'Go to crater X by way of outcrop Y and make sure you call at valley Z.' The controller would be confident that the hardware would do the job, but would leave the decision as to how it is done to the device's inbuilt intelligence.

There are several difficult problems to overcome in designing telepresence units for outer space. First, the machine must have very sensitive 'actuators' – mechanical hands or fingers or transport mechanisms – that will do jobs as well as humans.

Second, efficient communications links must connect the oper-

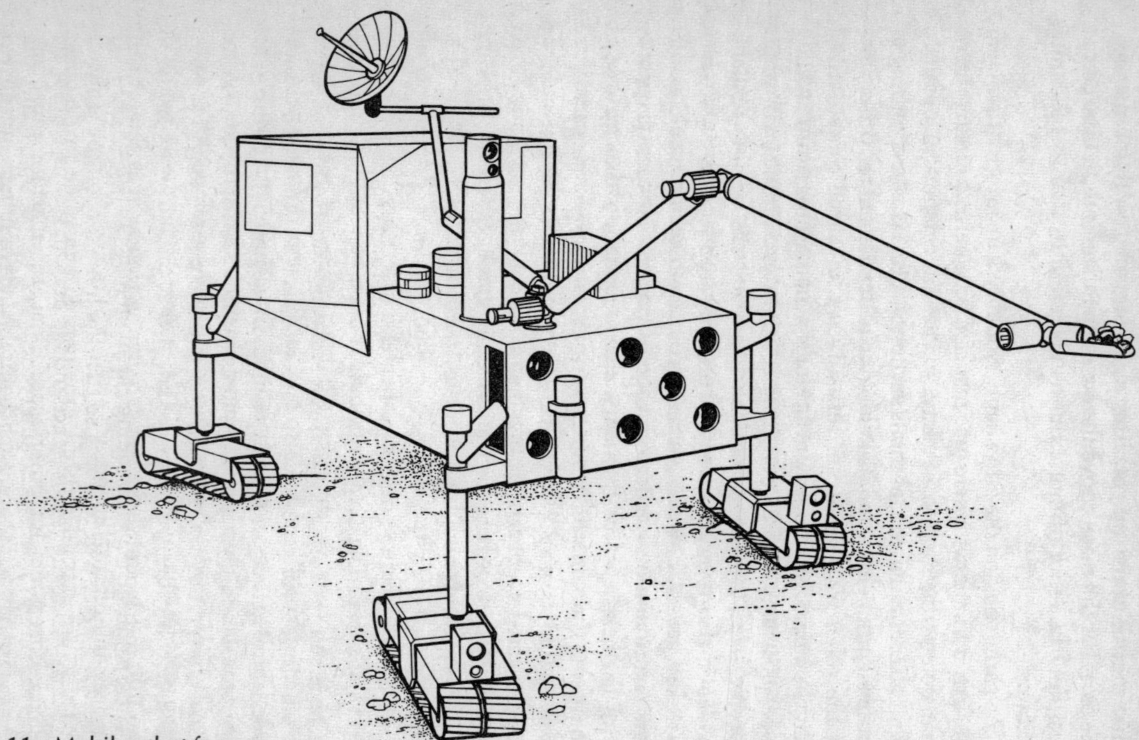

Figure 11. Mobile robot for space

ator with the unit itself. For operations deep into space, the links would probably involve satellites in geostationary orbit acting as bridges between points above the atmosphere and ground stations. Examples of such vehicles are those in NASA's system of tracking and data relay satellites.

The third tricky area is to supply the hardware with the essentially human expertise to do jobs that humans manage without any trouble, but which are, in reality, extremely complicated. These tasks include handling operations such as picking up a knife and fork or combing one's hair. Such skills, which can be managed with little difficulty by a child of six, are very difficult to 'teach', at least with today's technology, to machines. This is where the machine would have to be programmed with highly advanced computer software to give it the kind of general knowledge about simple tasks that is present in most people's brains. As an illustration, with a telepresence system for space repairs, the operator should not have to think about how his hand holds and positions a screwdriver before giving the unit instructions to do the same thing. Instead, he would simply tell the hardware, perhaps by typing on a keyboard, 'Put screw A into hole B'. A system operating in this fashion is, in the technical jargon, 'transparent'.

As the MIT workers point out in their report:

> From a human factors viewpoint, the more transparent the system, the better for operator performance. What this means is that the operator should not be performing tasks which would not have to be done if the operator were actually at the worksite. A mechanic rarely consciously considers how to position and move his arm while working; the control of his arm is automatic. If he were forced to plan and think about each movement of his arm, his work would be slowed down. A more extreme example is that of a runner. If a runner were to try to control his legs by actually thinking about the kinematics of motion, he would probably not be able to walk, let alone run.

The final difficulty in building the man–machine combination is to design a system of feedback so that the human operator is aware of exactly what the telepresence unit is doing. This is so that he knows enough to be able to intervene when necessary, in just the same way that he might control the movement of his arm or leg. TV cameras would have to send pictures of events taken from several different angles to the supervisor who would probably obtain different pictures by inclining his head in different directions, in the same way that, under normal circumstances, people change the view they get of the world. The pictures would very likely appear in a screen set into the operator's helmet.

Another technique envisaged is that the person in charge of the machine could command it by voice control to give him different pictures. This would require a computer to process the sounds of the human voice and obtain an interpretation of what someone is saying. The computer would then pass on the information to the mechanism controlling the flow of TV pictures.

According to the MIT workers, NASA should immediately start to plan for telepresence operations. This would mean tackling the various areas of research, ready to assemble and fly a prototype telepresence unit in 1992. Almost certainly, the first such system would be used for the servicing of spacecraft. Increasingly, space engineers are planning to repair or refuel (with the propellant to keep craft in the right orbit) space vehicles while they are still at work out of the atmosphere. In this way, the life of what can be extremely expensive chunks of hardware can be prolonged. This applies, in particular, to costly and prestigious scientific observatories placed in space.

One such vehicle is the $1,000 million Space Telescope, a joint venture between NASA and the European Space Agency. The Space Telescope is due to enter orbit in 1986. It will look at distant galaxies with accuracies far greater than are possible on Earth; because the instrument is above the atmosphere, it will collect radiation in a reasonably pure state, unattenuated by the gaseous mantle that lies between us and the rest of the universe.

This instrument is probably the most exciting scientific space project of the 1980s. It will pick out stars and other objects fifty times fainter than are possible to detect with ground-based instruments. Whereas the best Earth telescopes can 'see' 2 billion light years (2×10^{22} km) into space, the Space Telescope will look much deeper – to 14 billion light years. In other words, it will see events as they were 14 billion years ago, about the time the galaxies were formed. Readings from the telescope may, therefore, give scientists new clues about the nature of the 'big bang', the event thought to have created the universe.

With a device of this importance, NASA and ESA not surprisingly want it to remain in orbit for as long as possible. So the two administrations plan regularly to service the hardware rather as cars receive periodic checks to keep them in good condition. In this way, the life of the telescope may be increased from ten years to more than fifteen. Initially, teams of astronauts will do the work. Armed with special tools, these space-age equivalents of grease monkeys will be deployed from a space shuttle and zoom about with what are called manned manoeuvring units, small boosters attached to back packs. These space workers will replace small modules of the spacecraft before they have a chance to wear out. They will also attend to any faults. Later, however, the astronauts may be phased out in favour of telepresence units of the kind that the MIT workers envisage.

The telepresence units would be evolutionary advances on one piece of hardware that has already seen service in space, the shuttle's remote-manipulator system. This is a 15-metre arm, built by Spar Aerospace of Canada, anchored in the shuttle's payload bay. Controlled by astronauts in the vehicle's crew quarters, the arm can pluck satellites from orbit and bring them back to the cargo bay. Later versions will probably contain grappling units with telepresence so that they can reach out from the USA's space plane to repair spacecraft without direct human intervention.

The Canadian-built arm played a large part in a historic occasion in April 1984 – the first repair of a satellite while in orbit. The maintenance job, accomplished by two astronauts from the crew of a space shuttle, seems certain to be the forerunner of many others. The ailing satellite was the Solar Maximum Mission, a scientific vehicle to take measurements of the Sun, which was launched in February 1980 but failed just nine months later. The problem was faulty components inside the spacecraft's electrical equipment. As the vehicle remained roughly in its correct orbit about 500 km from Earth, an orbit within easy reach of the space shuttle, NASA decided to attempt a repair during the eleventh shuttle mission. The venture started badly after George Nelson, one of the crew of five, failed to attach a grappling device to the satellite, which was spinning slowly in space. Nelson, with a manned manoeuvring unit attached to his back, travelled through space for a short distance from the shuttle and located the stricken space vehicle. But, as a result of his own momentum, he succeeded only in pushing the satellite further out of control and had to retire. On 10 April, the crew succeeded, after Bob Crippen, the commander, steered Challenger close to the Solar Max and yanked the spacecraft into the payload bay with the aid of the Canadian arm. Ground staff at NASA's Goddard Space Flight Center had earlier slowed down the spin of the satellite by sending radio signals to it. The next day Nelson, together with James van Hoften, entered the payload bay in space suits to replace a box of electronics and repair another electrical fault. The latter job was quite fiddly and involved tasks such as untightening screws and rewiring. The complete exercise took a little more than seven hours, the second longest spell US astronauts have spent outside their vehicle in space. (The record was a 7.5 hour Moon walk during the Apollo 17 mission.) Later in 1984, during the fourteenth shuttle mission in November, two further astronauts carried out a similar exercise. Joseph Allen and Dale Gardner 'rescued' two satellites, Westar 6 and Palapa B2, that an earlier shuttle flight the previous February had placed in the wrong orbits. The satellites were returned to earth for repair and resale.

The American aerospace firm Grumman is already working on devices similar to the remote-manipulator system which would be used by humans in further in-orbit repair missions. These include platforms joined to the shuttle on which astronauts would stand to do intricate jobs on spacecraft. Systems such as these could also be adapted to include telepresence. CNES, the French national space agency, is also interested in the concept of robot repair vehicles. These could form part of an unmanned space station called Solaris that the agency is planning for the 1990s.

Besides the Space Telescope, other scientific vehicles likely to receive regular checks by human or mechanical maintenance teams include several American scientific spacecraft planned for later this century. These are: the gamma-ray observatory, scheduled for launch in 1988; the advanced X-ray astrophysics facility, due to enter space in 1990; and three spacecraft still on the drawing board for the 1990s, the very large space telescope (an instrument of 22 tonnes), the 67-tonne coherent optical system of modular imaging collectors (COSMIC) and the 85-tonne thinned-aperture telescope.

These last three vehicles are so big that they will probably have to be joined together in orbit from modules taken into space separately. Telepresence units could play a big part in such assembly operations. This kind of work, in constructing the elements of space stations for example, would be too difficult and dangerous for people. Space robots could also lend a hand in operating the orbiting factories that seem likely to appear next century. Humans are unlikely to volunteer for such activities, in the same way as, on Earth, people hardly queue up for jobs (unless they are paid fantastic wages) in inhospitable places such as the Antarctic or on North Sea oil rigs.

Telepresence hardware will probably feature in future missions to the planets. Of course, researchers have always relied on machines to do this kind of work for them. This is for the straightforward reason that, particularly with conventional rocket technology, to put people on space vehicles journeying far out into the solar system would be incredibly dangerous. Most of the unmanned probes and landers that the USA and USSR have launched have had very little 'intelligence' of their own (Table 17). They have relied generally on fixed programs of instructions, for example on what to do when they get to certain points during a 'fly-by' of a planet, or on the route to take in getting there in the first place. At best, these instructions can be changed only intermittently or with little exactitude.

A big problem with conventional probes is the time lag in transmitting signals between the spacecraft and the control station on Earth. These signals are sent by radio waves, which have a finite velocity. The further the distance, the more noticeable is the lag. Thus people on Earth who make international telephone calls by satellite hear the sound of the other person's voice a split second after he has spoken. This is because the signal has to bounce 36,000 km up to the satellite and back again to Earth. With transmissions to the Moon, the time lag is a few seconds, which is still only a minor problem. However, for any communication with a space probe on Mars, a controller on Earth would have to wait about forty minutes to determine whether his instructions have been acted upon. This is the time between sending the signal and receiving one back from the surface of Mars. With probes further out into the solar system, this time lag is counted in hours.

With the technology available today, therefore, space controllers are unable to supervise missions to faraway places with any certainty or degree of accuracy. Often, it is useless for a spacecraft in trouble to radio to Earth for an instruction, as the message would arrive too late to avoid a catastrophe. This increases the need for telepresence systems with some degree of inbuilt decision-making capability. Ground controllers could then send the most important messages, about the key areas that the probes should explore for

Table 17. Major landmarks in unmanned exploration of the Solar System

1959	Luna 1 (USSR)	first Earth-escape spacecraft
1959	Luna 2 (USSR)	impacts on Moon w3m1959
	Luna 3 (USSR)	returns TV pictures from far side of Moon
1962	Mariner 2 (US)	flies past Venus
1963	Mars 1 (USSR)	flies past Mars
1964	Ranger 7 (US)	returns close-up lunar pictures
1965	Mariner 4 (US)	sends back to Earth photos of Mars
1966	Luna 9 (USSR)	soft-lands on Moon
1966	Luna 10 (USSR)	enters lunar orbit
1966	Venera 3 (USSR)	impacts on Venus
1970	Luna 17 or Lunokhod 1 (USSR)	moves over Moon's surface in response to controls from Earth
1970	Venera 7 (USSR)	soft-lands on Venus and returns radio signals
1971	Mars 2 (USSR)	impacts on Mars
1971	Mars 3 (USSR)	soft-lands on Mars
1971	Mariner 9 (US)	enters orbit around Mars and takes pictures
1972	Venera 8 (USSR)	analyses soil samples from Venus
1973	Pioneer 10 (US)	returns pictures from Jupiter
1974	Mariner 10 (US)	sends data and pictures from Venus
1974	Mariner 10 (US)	flies past Mercury
1976	Viking 1 (US)	soft-lands on Mars and analyses soil
1977	Venera 9 (USSR)	sends to Earth TV pictures from surface of Venus
1979	Pioneer 11 (US)	returns pictures of Saturn
1979	Voyager 1 (US)	sends pictures of Jupiter and moons
1980	Voyager 1 (US)	sends pictures of Saturn and moons
1983	Pioneer 10 (US)	leaves solar system after journey of 5.6 billion km

example, and leave the minute-by-minute judgements to the control circuitry built into the device's hardware.

Soviet space planners know only too well the difficulties of attempting to control every aspect of an unmanned mission of exploration from far away. In 1970 and 1973, the USSR put onto the surface of the Moon two unmanned devices which responded to signals from the Earth. These Lunokhod machines, which resembled Heath Robinson inventions, had wheels so they could trundle over interesting parts of the lunar landscape. In the two missions, the vehicles clocked up a total of 50 km. They moved not independently but under the control of a team of technicians at the USSR's space centre near Moscow. Looking at the TV pictures that the vehicle supplied of its surroundings, the technicians had to steer it around obstacles such as rocks, always bearing in mind the delay of several seconds between the movement of the machine and the reception of pictures on Earth. Even with this fairly small time delay, the problems were horrendous. According to reports in the West, the mental strain brought several members of the supervisory team close to breakdown.

A mission of an unmanned vehicle into the solar system with some degree of inbuilt intelligence is unlikely before the year 2000. The planetary projects due to go ahead over the next few years are mainly variations on previous missions and provide the probes themselves with little or no independence of ground controllers (Table 18).

Table 18. Planned unmanned probes into the Solar System

1986	Four spacecraft (two from Soviet Union, one each from European Space Agency and Japan) due to encounter Halley's Comet
1986	Galileo mission to Jupiter (US), due to arrive 1988
1986	International Solar Polar Mission (ESA and US); spacecraft will fly over pole of Sun
1986	Possible Soviet mission to Phobos, moon of Mars
1988	Venus radar mapper (US)
1988	AGORA, possible ESA mission to asteroids
1989	Possible French/Soviet mission to Venus
1990	Mars orbiter (US); arrive 1991
1990	Possible launch of US probe to comets
1993	Possible US probe to study Titan, moon of Saturn

A NASA study in 1980, however, pointed the way towards space exploration using vehicles that make decisions. According to the study, a probe would be sent on its way to Titan, the largest of Saturn's moons, around the year 2000. Researchers drew up a list of capabilities that the craft would require if it were to navigate to the moon, land, survey the terrain and return information to Earth. The vehicle would have to select interesting sites for reconnaissance; plan the different stages of the mission, including the strategies for deploying landing probes; navigate by planning trajectories; maintain communications links and thermal control; budget for energy requirements; diagnose faults; co-ordinate and control sensors; handle and analyse samples; organize and reduce data from sensors to a level that Earth control would be able to manage; and generate and test hypotheses, for instance about reasons why aspects of the mission do not work as planned.

Perhaps the most interesting vision was provided by Georg von Tiesenhausen and Wesley Darbro, two engineers at NASA's Marshall Space Flight Center. For some years, space planners have talked about establishing bases on the Moon as a follow-up to the Apollo missions of exploration. Von Tiesenhausen and Darbro took these ideas a stage further with a blueprint for how robotic devices could operate Moon factories. Built into the Moon base would be mechanisms to ensure that the manufacturing plant never runs out of workers – the robots would self-replicate.

A Moon factory appears attractive mainly because our nearest neighbour in space is rich in many of the raw materials that we need on Earth. The Moon contains plenty of oxygen, silicon, iron, aluminium, titanium and magnesium, substances which are important in the construction and manufacturing industries. Mines on the Moon to produce these substances, so the argument goes, would supplement the rapidly depleting stocks of such materials on Earth. Having obtained the raw material, it would not make sense to transport the substances back to Earth for processing. A more promising notion is either to take the riches from the Moon half-way to Earth, to the orbiting factories discussed in the last chapter, or to have the manufacturing plants adjacent to the mines on the lunar surface. Working on the Moon would hardly appeal to humans; hence the plan to employ teams of robot labourers.

The ideas of von Tiesenhausen and Darbro, sketched out in a

NASA report in 1980, follow on from pioneering theories on automation worked out by the famous mathematician John von Neumann. In 1952, von Neumann, a Hungarian–American who had already given to the world several important guidelines on how to build and apply electronic computers, started work on a book on automata. After the great man died, Arthur Burks, a colleague, edited the unfinished manuscript and the volume was eventually published in 1966 under the title *Theory of Self-Reproducing Automata*.

Von Neumann postulated that a self-reproducing mechanism must have four components. Component W is an automatic factory that collects raw materials and turns them into products according to a written instruction. Component X is a duplicator which takes the set of instructions and copies it. Component Y is a controller, linked to W and X. When Y is given an instruction, it first passes it to X for duplication, then to W for action and finally supplies the copied instruction to the parts that W produces, so that the whole operation can begin again. Component Z is a written instruction which contains the specification to let W manufacture the complete system – a new W plus X plus Y.

The mechanism proposed by the NASA engineers would follow these principles (Figure 12). According to von Tiesenhausen, assistant director of advanced projects at the Marshall Center, the mechanism would turn out not just new robot workers but brand new factories that would mushroom over the lunar surface at an exponential rate (Figure 13). The first factory on the Moon would probably take about a year to build. It would take another twelve months to produce its first offspring – a replicate factory. After a decade, there would be about 1,000 of the plants – and 1,000 million after thirty years. Of course, engineers would be faced with the problem of telling the self-reproducing machinery to stop well before this point was reached, otherwise the rapidly expanding manufacturing workshops would run out of space and be forced to start creating new factories out of other parts of the Solar System.

Von Tiesenhausen says that the ideas for the reproducing factories have their serious side. The concept provides a good way of expanding output from the Moon with only minimal engineering effort. Von Tiesenhausen is interesting not only for his espousal of the thought-provoking principles of robots that have 'children', but

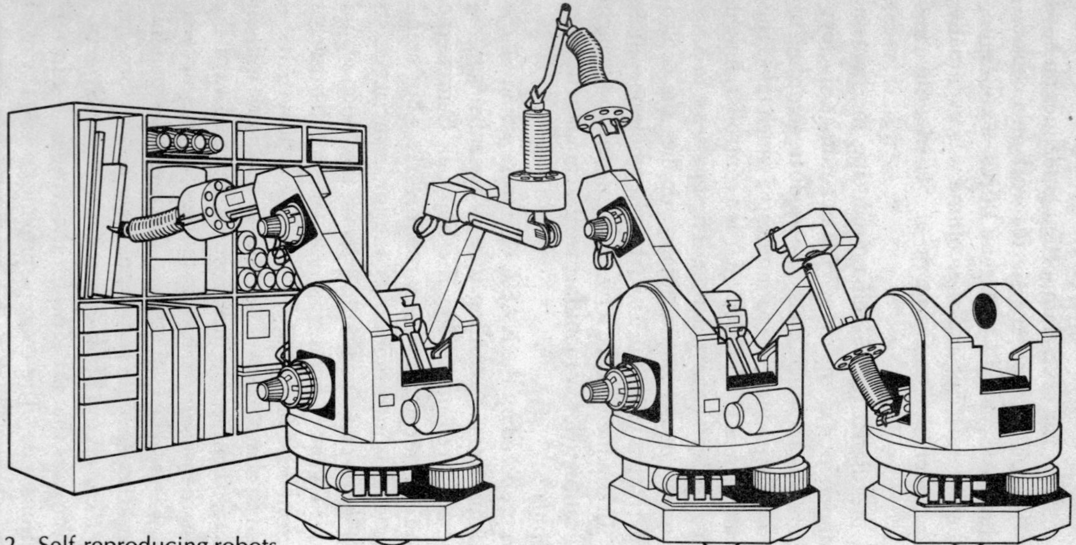

Figure 12. Self-reproducing robots

Replicating systems
Demonstration program summary
Object: automated robot manufacture capability
Phase A: Duplication of robot assembler by robot assembler, using supplied parts
Phase B: Performance of useful tasks and/or creation of useful products by robot assemblers
Phase C: Duplication of robot assembler from feedstock
Phase D: Duplication of robot assembler from raw materials

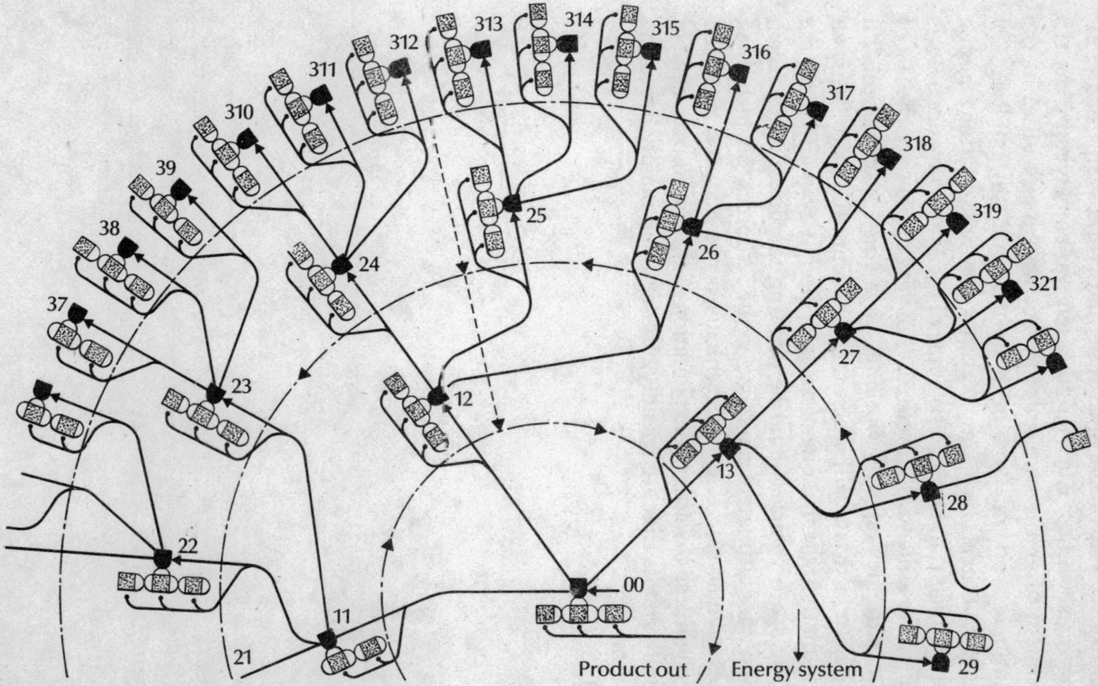

Figure 13. Growth plan of Moon factories

also for his being a former Peenemünde engineer who worked on Germany's V2 weapons during the Second World War (Chapter 2). Now in his seventies, von Tiesenhausen was employed by the US government after the war ended, along with many other of the Peenemünde workers. Now he is the last one still employed by NASA; the others, including Wernher von Braun (after whom a shopping precinct in Huntsville, home of the Marshall space base, is named), have either died or retired. 'I'm very proud to be the last of the Peenemünde staff still working,' says von Tiesenhausen. 'I'll still be here [at Marshall] when they condemn the place.' It is an interesting thought that one of the men directly responsible for the first true space vehicles could be pointing the direction in which mankind may ultimately capitalize on the resources to which these vehicles give access. If von Tiesenhausen has his way, men and women will step towards the exploitation of these resources hand in hand with robots that are friendly, obedient and rapidly becoming more numerous.

9 ☆ Rockets: Technology and Competition

The space ships of the next century will be powered by a range of exotic technologies, some of which will bear little resemblance to the way today's rockets obtain their energy. Over the coming decades, designers will seek to improve upon some of the inherent inadequacies of the mechanism that has featured in most of the applications of space to date – the chemical rocket. This means of propulsion has existed in its most useful, liquid-fuelled, form for a little over fifty years. The rocket's weaknesses, however, were spotted early on by space planners. Some of these inadequacies will be remedied in the rockets that will figure in the early part of the twenty-first century, which will be powered, for at least some of their journeys, by nuclear energy, streams of ions or the flow of photons irradiated by the Sun.

The chemical rocket has served the community of space engineers well. It has provided a reasonably straightforward way of blasting small payloads a short distance into the sky against the extremely powerful force of gravity. Most of these payloads have been satellites, which rarely weigh more than a meagre couple of tonnes. What is more, the period of 'burn' during which rocket engines fire rarely lasts more than a few minutes. When it comes to boosting very large structures away from the Earth, for instance the space platforms that are now on the drawing boards whose masses may run to hundreds or even thousands of tonnes, the chemical rocket is by no means as useful.

The mechanism's inadequacies would become equally apparent a long way out into the solar system, far away from the areas where the gravitational tug from planets is highest. On these missions, a substantial burn time to produce a high acceleration would be required if the rocket's journey were not to take an impossibly long time.

A rocket's main job is to provide a large thrust. This thrust produces a reaction in the opposite direction, which pushes the rocket against the force of gravity and away from the Earth. This much is predicted by Newton's Third Law of Motion. With the 'push' generated by its engines, a rocket attains the high velocities required to put its payload either in orbit or in a parabolic path that takes the object away completely from the gravitational attraction of the Earth and out into the solar system. The minimum velocities required to place payloads into these two trajectories are 8 km/s and 11.2 km/s (see Chapter 3).

Chemical rockets provide a lot of thrust for a short time. This is useful for space engineers preoccupied with combating the immense force of gravity close to the Earth's surface. Gravitational attraction falls off rapidly the further an object travels from the planet. The force, in fact, diminishes in inverse proportion to the square of the distance from the Earth. Thus a rocket in the early stages of its ascent requires a tremendous amount of energy to wrest itself from the gravitational pull. This is emphasized in the main engines of the USA's space shuttle, which burn for only eight minutes to lift the rocket off its launch pad and away from the Earth. During this time the three engines burn 2 million litres of combustants in the form of liquid oxygen and liquid hydrogen. The total force generated by the shuttle at lift-off is 3 million kg. After all this effort has been expended in lifting the vehicle and its payload a couple of hundred kilometres above the Earth, it is a much easier job to send any satellites on board further from the planet. This is done with booster rockets (called upper stages) that are fitted to the satellites themselves inside the shuttle's cargo bay.

In chemical rockets, a thermodynamic reaction takes place between substances which burn in a combustion chamber. The substances are a fuel plus an oxidizing agent producing hot exhaust gases which stream out of the engine's nozzles. The hotter these gases are, the faster they travel. A rocket's velocity is proportional to the speed at which the exhaust gases eject from the nozzles of its engines. So designers who want high-performance rockets must make their chemical reactions as hot and as furious as possible. However, chemical reactions, even the most vigorous ones, cannot attain temperatures above a few thousand degrees centigrade. This limits the velocity at which the combusted gases stream out of the exhaust. It is impossible, with a simple chemical rocket, to attain a forward velocity of more than about 3 km/s. Space technologists can gain higher speeds with the principle of staging. In this several independent rockets, each with their own combustion chambers and fuels, are stacked on top of each other. Successive stages start to burn when they are already travelling at a high speed. With staging, the final velocity of the hardware can reach 12 km/s, enough, for instance, to send a space capsule on its way to the Moon. Even with this kind of technical stratagem, chemical propulsion cannot provide rocket velocities above about 25 km/s. This is far below the kind of velocities which would be required for very long trips far into the solar system and beyond.

Another way in which the velocity of chemical rockets can be increased is to eject a greater mass of gas from the exhaust, but for this to happen the fuel that the rocket carries must also increase, adding to the total weight that must be pushed away from the ground against gravity. So increasing the fuel is, by itself, a wasteful and inefficient way of increasing velocity.

Space engineers try to keep the total mass of the craft to a minimum by ensuring that the fabric of a rocket is very light. In modern rockets, some 95 per cent of the vehicle's lift-off weight is accounted for by fuel. (In the case of the German V2 rocket of the 1940s, the figure was 70 per cent.) With chemical propulsion, a rocket of total mass of some 80 tonnes (most of this fuel) is needed to lift a satellite of a mere 200 kg. For larger payloads, the overall weight of the vehicle increases still more, placing an inherent limitation on the loads that chemical rockets can carry. The most powerful rocket yet produced was the Saturn vehicle which sent US astronauts to the Moon. This could place a mass of some 100 tonnes into a low orbit 300–400 km from the Earth.

A rocket's ability to turn a given amount of propellant into thrust is described in terms of a quantity called specific impulse. The specific impulse of a rocket is the thrust produced per unit weight of propellant consumed in one second. It is a measure of the vehicle's capabilities, analogous to the figure for a car giving the number of kilometres it will travel on a litre of petrol. Chemical rockets have fairly low specific impulses. Either they provide a reasonable thrust over a short time, or they give a puny force over a longer period. Either way, chemical rockets are not suitable for some of the jobs that space planners think are ahead.

Specific impulse is measured in seconds. Chemical rockets are limited to specific impulses of no more than about 500 seconds. The Centaur rocket to be used as an upper stage on the space shuttle stretches chemical rocket technology to its limits. The rocket has a specific impulse of 446 seconds. Earlier generations of rockets based on the chemical principle have lower specific impulses. The figure for Blue Streak, Britain's launcher of the 1960s (of which more later in the chapter), was 248 seconds.

There are several ways in which specific impulse can be increased, to levels well beyond that possible for chemical rockets. One technique is to use the energy generated by nuclear reactions. A fission reaction of the kind that occurs in nuclear power stations produces much hotter conditions than are attainable with chemical reactions. Temperatures of hundreds of thousands or even millions of degrees Centigrade are, in theory, possible. The heat from the fission is passed to a working fluid such as hydrogen gas, which provides the rocket's thrust by streaming out from the exhaust at a high speed.

Still more useful rockets may be possible with nuclear fusion, the mechanism that takes place naturally in stars such as our Sun. In fusion, two light elements join together to provide a great outburst of energy, which can again be passed on to a working fluid. Alternatively, this energy can produce the rocket's explosive thrust directly. Fission systems promise specific impulses of up to 1,000 seconds; with fusion, the figure could be up to 100,000 seconds. Using fusion propulsion, we can obtain an exhaust velocity not of a few kilometres a second but of the order of a few thousand kilometres a second. Such rockets may be required if mankind is ever to journey to even nearby stars, which are at incredible distances from Earth. The distances of such stars are reckoned in terms of light years, one light year being about ten million million kilometres – the distance travelled by electromagnetic radiation in one year.

There are plenty of practical problems in constructing rockets that work according to nuclear principles. In practice, the fission process is easier to achieve. It requires only a certain amount of nuclear fuel, the critical mass, to make the reaction occur unaided. Fusion, however, provides more energy for a given amount of mass and so appears the more attractive option for space flights, particularly for very long journeys to distant stars which we may want to make in the future. However, to make nuclear fusion work, it is necessary to heat pellets of the atoms to be fused to very high temperatures, to hundreds of millions of degrees Centigrade. One way is to use a fission reaction as a 'heater'. This is the principle employed in the hydrogen bomb, a horrific marriage of fission and fusion technologies. The volumes of energy released in this way, however, would be too large for what is required; and, being completely uncontrolled, such reactions could destroy not just the rocket and its crew and cargo, but anything else that happened to be in the vicinity.

A promising fusion technique is to heat fuel in a more controlled way, with the localized energy produced by lasers, for example. Alternatively, the materials to be fused could be held in the form of a plasma, then heated by high-speed electrons. This is the concept adopted by engineers attempting to build the prototypes of fusion power stations, which, if they ever work, could solve many of the world's energy problems some time next century. It was also the approach chosen by scientists and engineers from the British Interplanetary Society, who between 1973 and 1977 conducted a theoretical study of how a fusion spacecraft would operate.

In the study, called Project Daedalus, the group proposed a fusion engine which produced explosions at the rate of 250 a second, enough to drive a spacecraft with fantastic acceleration. The power developed by the vehicle would be several times that generated by the entire Earth today. There would be no way of stopping the vehicle; near its destination (perhaps a star several hundred million million kilometres away), it would have to release subsidiary spacecraft to relay information back to Earth.

There is also the small matter of the fuel that the rocket would have to carry. According to the Project Daedalus study team, the fuel would be pellets of deuterium (an isotope of hydrogen) and a light isotope of helium. Some 30,000 tonnes of the helium isotope would be needed, which would pose special problems as it is virtually unobtainable on Earth and we would have to travel to Jupiter to find a sufficient quantity!

Clearly this discussion is moving into the realms of fantasy. Yet the past few decades have seen several serious efforts to produce nuclear devices for space vehicles. The work has concentrated on nuclear fission, this being a technology that at least is fairly well established in terrestrial applications.

In 1955, the US Atomic Energy Commission started work on a programme known as Space Nuclear Auxiliary Power (SNAP). This was intended to provide a new source of electricity, up to 50 kW, for space vehicles. Other efforts to produce nuclear reactors for space have continued. Nuclear energy has featured in many of the probes and satellites sent by the USA and the Soviet Union either around the Earth or to distant planets. The craft have been powered by small reactors or by the heat produced by the decay of nuclear isotopes.

During the 1960s and early 1970s, government agencies in the USA pushed ahead with work to channel nuclear fission specifically to new forms of rocket engines. These projects, none of which produced anything workable, went under names such as Nerva, Kiwi and Phoebus.

In the early 1980s, the notion of building nuclear reactors in space attained a new lease of life in the USA. This was due mainly to enthusiasm by members of the US defence research community, who want nuclear devices not so much to power engines as to supply large amounts of electricity to weapons systems such as lasers (see Chapter 6) or huge orbiting radars. Whereas most satellites require power of no more than a few kilowatts (which is easily obtainable from solar panels), space hardware of the kind envisaged by the military would be driven by hundreds of kilowatts, which nuclear energy could provide.

If the day of the nuclear rockets is still in the future, the same can be said of the other unorthodox forms of propulsion system that space theorists think could improve on the chemical formula. A promising technique is to accelerate ions by an electrical field and push them out of a rocket nozzle. Such systems could have specific impulses of between 1,000 and 10,000 seconds. They would yield a low thrust but over a long period, making the rockets useful in manoeuvres away from the biggest attractive forces of the Earth's gravitational field and where long burn times are required.

So far, ion engines have powered only small space vehicles. In 1964, NASA tested such a system in an experiment called SERT (Space Electric Rocket Test) that lasted just fifty minutes. In this, engineers launched a spacecraft which manoeuvred in orbit with engines that push out ions of mercury. This was followed up by a much more extensive programme which lasted eleven years. During what was called the SERT-2 project, a spacecraft powered by similar engines was controlled in a polar orbit. In the engines, mercury atoms are bombarded with free electrons, which knock out electrons from the atoms to produce positively charged ions. These ions mix with other electrons to produce a hot plasma; a magnetic field draws out the ions from the plasma and they exit at high speeds to provide the rocket's thrust.

The main application for these and other novel forms of rocket is to lift objects from fairly near to the Earth to points further away. For example, ion-beam or nuclear rockets could transfer satellites or other structures from the low Earth orbits, a few hundred kilometres from the planet, to the geostationary circle 36,000 km away. Ordinary chemical launchers would initially lift the load to the low orbit.

A promising candidate for a spacecraft powered by ion beams is a vehicle which may mount an ambitious mission to the asteroids in the 1990s. The voyage is one of several exploits in space science that a team of researchers from NASA and the European Space Agency is considering. The 500 million km journey to the asteroids, which are small lumps of rock lying mainly in a belt between Mars and Jupiter, could be made by a spacecraft powered by a West German ion engine.

The preference for this kind of rocket motor is due to the high power levels that engineers think the voyage would require. The spacecraft would need a powerful propulsion system not so much as a result of the long distance it would have to travel but because of the complicated changes in direction that the vehicle would be required to make. The mission would not be a simple 'fly-by' of a single planet but would rather involve the spacecraft in weaving in and out of orbit around a number of individual asteroids, which are generally a few hundred kilometres in diameter. This kind of trajectory would impose a colossal drain on the propulsion unit, requiring a power level of about 25 kW, compared to the few hundred milliwatts required by a fly-by probe of the kind that has been sent to Saturn or Jupiter. To develop this kind of power in a chemical rocket over a long period would require an immense amount of fuel. On the other hand, it has been calculated that an ion-engine motor suitable for such a voyage would require, at the start of its journey, just 1 tonne of propellant in the form of liquid mercury. The fuel would account for around half the weight of the spacecraft on the launch pad.

In another propulsion technique, called solar sailing, vehicles would simply drift through space powered by the propulsive force of photons streaming out from the Sun. Photons (invisible particles which give rise to light) exert a tiny though useful force when pushing against a large 'sail' of gossamer material in the vacuum of space. Amateur groups in the USA, France and Britain are interested in building space vehicles which use solar sailing. These craft, which would, like sailing boats, 'tack' to travel in directions other than directly away from the Sun, could be in orbit in the late 1990s. The concept of solar sailing appeals to those who dislike the expensive, high-technology aura of many space projects. Solar sailing, by contrast, would be a cheap way of travelling, using an energy source that is absolutely free.

Over the next few years, however, ideas about novel propulsion systems will be far from the thoughts of those concerned with putting satellites into orbit. Chemical rockets will play the leading role in space-transportation technology for the remaining years of this century, if only because alternative mechanisms are expensive and extremely difficult to produce. Consider, for example, the most grandiose space project to be mooted in the past few years. This is a scheme for a series of solar-powered satellites that would girdle the Earth to provide masses of cheap energy. The satellites, first proposed by an American engineering consultant, Peter Glaser, in 1969, would be vast affairs. Each would measure some 10 km by 50 km, about the size of Manhattan Island. These huge platforms would be coated with solar cells which would convert rays from the Sun to electricity. The electricity would be converted to microwaves and beamed to Earth, where it would be collected by large receiving antennae, perhaps built on artificial islands. A network of 100 satellites in the geostationary orbit would provide 500 GW of energy, equivalent to the output of several hundred nuclear power stations.

How, though, would the material for such immense structures be lifted into orbit? Each satellite would weigh about 50,000 tonnes and cost billions of dollars. In conjunction with NASA, Boeing in the USA has produced a design for a large space freighter which, though visionary in many aspects, features tried and tested chemical-rocket technology. Boeing's heavy-lift launch vehicle, to give the craft its full name, would look rather like two space shuttles, one on top of the other. Both segments of the craft would have wings and be re-usable. Each of the 'shuttles' would be twice the size of today's version of the space plane, giving the complete heavy-lift launch vehicle a height of 170 metres. Both the top and the bottom segments of the spacecraft would fly back to Earth after the top section had delivered, to low orbit, its payload of some 400 tonnes (more than ten times the capacity of the 1980s version of the shuttle). The whole vehicle would weigh 10,000 to 20,000 tonnes, compared with 2,000 tonnes for today's space plane.

According to the space planners, parts of the solar-powered satellites would be constructed in the low orbit a few hundred kilometres above the Earth. They would then be transported to their final resting place in the geostationary ring by other kinds of spacecraft. Here less conventional technologies could come into play; for example, these craft could use propulsion systems based on ion engines. Looked at in the sobering context of the economic conditions of the 1980s, the cost and complexity of a system of such satellites appear immense. This makes it unlikely that the system will be built in the lifetimes of anyone living today.

Any such network of space structures could be constructed, almost certainly, only as a result of some great initiative by the world's governments. This follows the pattern of the past couple of decades in which space activities in general have been dominated by government bodies, mainly of the USA and the USSR. (Other countries which have put objects into space with their own rockets are France, Britain, Japan, India and China. Pakistan and Brazil may soon join this élite club of nations.) Putting hardware into space has certainly not been thought of as a commercial enterprise in which private companies participate. But this state of affairs is changing, as private firms enter the arena of what has become a mature technology, and as the increasing number of countries and organizations who want to launch satellites start to shop around for the best deals.

The biggest competitors in satellite launchers are the Ariane rocket, developed by the European Space Agency and sold by a French-controlled company called Arianespace, and the American space shuttle. The two vehicles are contenders mainly for taking satellites for communication and other purposes into the geostationary orbit.

The historical background to Ariane almost deserves a chapter to itself. It arose from the desire in Western Europe during the 1960s and 1970s to establish a launch capability that would provide independence in this area from their ally the USA. The USA had given a few hints that, should the business of operating communications satellites ever turn into a big money spinner, then it might be reluctant to put into orbit space vehicles built in Europe on the grounds that these would compete with US craft. For this reason, some politicians and engineers in Western Europe felt it imperative that they should have a launch capacity of their own.

Initially, the leading light in this movement was Britain. The dominance that the UK enjoyed in Western European rocketry during the 1960s was a result of what, with hindsight, was a series of peculiarly muddled political decisions in the decade previously. In the aftermath of Britain's development, following the Second World War, of a capacity to make nuclear bombs, the country needed a way of delivering them. The most likely opponent in any future war was the Soviet Union. So to provide a nuclear deterrent, Britain had to demonstrate that it could drop a bomb on Moscow. Ground-launched nuclear-armed rockets were thought to be the answer. Though Britain had flirted with the idea of enticing some of the German V2 rocket engineers to its shores to work on missiles (see Chapter 2), it had little expertise in rocket engineering; but it did have a good base in aeroengines (in the shape of Rolls-Royce) and in the design of airframe parts (represented by De Havilland, a company since incorporated into British Aerospace). These two companies were the main contractors in the effort, to which the government agreed in 1955, to build a one-stage rocket called Blue Streak. (In the early stages, the project was top secret and labelled simply as Code 3000.) Rolls-Royce's job was to build the vehicle's two engines, fuelled by kerosene and liquid oxygen, while De Havilland had to build the fuel tanks and frame of the missile and was given overall responsibility for the project.

Ironically (in view of the concern about competition on a commercial level a decade or so later) the USA was anxious to help its ally to build up rocket technology. In an arrangement that was kept secret at the time, the US government provided Britain with design details of its own Atlas intercontinental missile. Convair and General Dynamics, the US contractors for the rocket, established links with their counterparts in Britain, Rolls-Royce and De Havilland. Though the UK firms did a lot of extensive redevelopment work, the basic shape and performance of Blue Streak was derived from Atlas.

To complement the rocket-building effort, most of which was at Rolls-Royce's engine plant at Derby and in De Havilland's Stevenage factory, a vast test site was built in desolate moorland in Spadeadam, Cumbria. The 400-hectare site featured engine-test stands 30 metres high and huge plants to turn out the thousands of litres of liquid oxygen and nitrogen needed for ground firings. (The nitrogen was required for purging of engines.) Wimpey, best known for its house-building activities, was put in charge of the construction effort. The site, used for thousands of ground firings between 1958, when it opened, and 1975, when the Blue Streak programme finally closed down, contained 10,000 km of electrical wiring – enough to stretch a quarter of the way around the world.

As it turned out, all that effort went to waste. By 1960, the government had had second thoughts about delivering nuclear bombs by means such as Blue Streak. The problem was that Britain would have had to build dozens of silos from which to launch the missile. Models of such installations, which would have been holes in the ground some 50 metres deep and surrounded by masses of concrete and steel, had already been tested. Government workers even started building one of them – in Wiltshire. Military tacticians, however, worked out that the silos would be vulnerable to pre-emptive attack by Soviet missiles, whose accuracy was improving all the time. In April 1960, the Macmillan government announced that it was abandoning Blue Streak as a military project. From this point, Britain put its faith concerning nuclear-bomb delivery systems either in aircraft such as the Vulcan or in Polaris, the submarine-launched missile purchased from America which was thought to be less vulnerable to attack than ground-based rockets.

With Blue Streak no longer needed as a military vehicle, ministers tried desperately to find another use for the rocket. Peter (now Lord) Thorneycroft, the then aviation minister, chaired a conference in Strasbourg in 1961 in a bid to persuade the rest of Europe to join a purely civilian launcher project. The effort touched a chord with other European governments mindful of the possibilities of satellite launches in the years to come.

In the agreement that emerged, Western European nations formed, in 1964, the European Launcher Development Organization (ELDO), whose task was to develop a three-stage launcher based on Blue Streak as the first stage. The other stages were to be Coralie, a booster provided by France, and Astris, West Germany's contribution to the project. In the sharing-out of spoils, Belgium opted to build the vehicle's guidance system and the Netherlands the communications hardware. Italy was to construct the satellite that Europa (the name given to the rocket) was intended to launch.

The programme soon encountered both technical and management problems. ELDO never succeeded in putting a satellite into the geostationary orbit, and the estimated costs of completing the programme climbed steeply, from £70 million in 1961 to £140 million in 1965 to £210 million in 1966. All this left Britain, which had agreed at the outset to finance 38 per cent of the Europa programme, wondering what it had let itself in for. It was left to Tony Benn, minister of technology in the Labour government, to announce in 1968 Britain's withdrawal from the organization that it had been instrumental in creating.

The Europa programme, however, staggered on, notwithstanding the lack of British involvement. Blue Streak was launched eleven times, either by itself or as part of a complete Europa rocket. The first launch was in June 1964 and the last in November 1971. The first ten firings were from a rocket range in Woomera in Australia (Australia was the seventh ELDO member), while the last took place from France's space centre in Kourou, French Guiana in South America. In each case, the technical record of Blue Streak itself was excellent, though the complete rocket suffered from several failures.

By the early 1970s, the Western European nations had become convinced that the management set-up of ELDO was a recipe for disaster. As Britain had relinquished responsibility for taking the technological lead, the organization was in effect run by a committee. In 1973, the ELDO participants decided to wind up the organization and abandon the Europa project. ELDO was merged with a second body, the European Space Research Organization, to form a new European Space Agency, which formally started up in

1975. Leadership in European rocket technology had by now passed from Britain to France, which was the driving force behind a new rocket programme for which ESA was to take responsibility. (The second big project of the new organization was Spacelab, in which Germany was the leading country.)

Initially, the new European launcher was called L3S. This stood for Lanceur Trois Étages de Substitution, or Replacement Three-Stage Launch Vehicle. Soon, however, the ESA members decided they wanted a less prosaic title. Various names were suggested. For some reason, many of them began with a P: Prometheus, Patience and Penelope. ESA was all set to call the rocket Vega – until someone pointed out this was the name of a French beer. Finally, the ESA delegates chose Ariane, perhaps because the rocket offered a route to salvation and they recalled that it had been Ariadne's thread which led Theseus out of the maze.

In all these discussions, Britain took a back seat. It was thoroughly disillusioned with rockets after its experiences with Blue Streak, so decided to contribute only about 2 per cent of Ariane's development costs – which by the early 1980s had totalled £700 million, 70 per cent provided by France. The UK decided that its main role in ESA would be in building satellites for communications, leaving others to shoulder the rocket burden.

It appears that the French decision to lead the Ariane development is paying off. The Centre Nationale d'Études Spatial (CNES), the French national space agency, has managed the technical effort and created a strong body of expertise in rocket engineering. Interestingly, top CNES officials say they learned much from the British efforts at Spadeadam during the Blue Streak era. The Ariane programme employs 2,500 people in France besides keeping going the French space centre in Kourou from which the rocket is launched. Arianespace, to which in 1984 ESA handed over responsibility for Ariane launches, is a semi-private body dominated by French interests. CNES (together with other, private, French companies) own about 70 per cent of the shares.

Not surprisingly, the Blue Streak saga has left bitterness among British rocket engineers. They feel that, with stronger government backing in the 1960s, Britain, not France, could have emerged as the number one rocket power in Western Europe. Britain has indeed obtained very little from the £600 million (1983 prices) it spent on

Blue Streak. The sentiments are summed up by Geoffrey Pardoe, who at the age of twenty-eight was the De Havilland engineer in charge of the Blue Streak project. He spoke for many Blue Streak engineers when he told me in 1984:

> The Blue Streak episode was a disastrous failure at top level to appreciate technological opportunities. The British were never totally determined about using Blue Streak to launch satellites. We suffered from a kind of schizophrenia – we had the capacity to build something which was not matched by the desire to do it.

At about the time the Europeans finally settled on the Ariane programme, the USA was starting the job of designing the space shuttle. Although both space vehicles use standard chemical propulsion technology, they are very different pieces of hardware. Ariane is an orthodox three-stage rocket. Such vehicles are called expendable launchers. They are destroyed in the process of taking into orbit their payloads, which sit inside a shroud in the rocket's nose.

The shuttle, on the other hand, is the world's first re-usable launcher.

By 1986, the USA plans to operate a fleet of four shuttles. Each comprises three parts: the orbiter or space plane itself, a manned vehicle about as big as a jumbo jet which is completely re-usable; the tank for fuel which is destroyed on each flight; and the two solid rocket boosters which are recovered after each launch and thus partly re-usable. The orbiter contains crew quarters, plus a cargo bay in which objects such as satellites are kept. The shuttle can take up to 30 tonnes into an orbit a few hundred kilometres above our planet.

The fact that many of the shuttle's components can be reflown time and again should, in theory, make the vehicle cheaper. A factor pushing costs in the opposite direction is that the US vehicle is a very versatile piece of machinery trying to be several different entities at once. Besides being a mechanism for putting communications satellites into the high geostationary orbit, it has an important role in placing spacecraft into low orbits. In fact, it is much better at this than as a carrier of geostationary vehicles. The shuttle itself can climb no higher than about 700 km from the Earth. One of the main reasons the vehicle was developed was to inject into space military satellites (spy craft and the like) which commonly use low orbits. Satellites required to enter higher orbits must be boosted into their trajectory by an upper-stage rocket inside the shuttle's cargo bay. A third role that the shuttle must play is as a carrier of people. The crew quarters contain room for up to eight men and women and several more could be squeezed, at a pinch, inside the cargo bay. The vehicle's developers reasoned that putting people into space along with hardware is often useful, and, as we have seen in previous chapters, it is. However, oxygen systems, waste disposal facilities and the other mechanical paraphernalia needed to keep people alive add greatly to the cost of what is, after all, the fairly simple job of placing satellites into space.

The extent of the rivalry between the shuttle and Ariane was shown in late 1983 when Britain finally had to make up its mind which vehicle would take into space the country's next generation of military communications satellites, the Skynet 4 series. Eventually, Ministry of Defence officials plumped for the shuttle. They thought it more reliable. The decision angered the French government, Ariane's main sponsor, and the episode revived memories of the grudging support for Ariane from Britain back in the 1970s when the programme was decided.

The competition between Arianespace and the operators of the shuttle has focused attention on the respective costs of the two spacecraft. When NASA virtually doubles its prices for the shuttle in October 1985, the cost to satellite operators of putting spacecraft into geostationary orbit will be roughly the same irrespective of whether they use the space shuttle or the European vehicle. At present (late 1984) the shuttle has a distinct price advantage. The cost of launching a satellite of about 1 tonne is $16 million for the shuttle, or two-thirds of the price for Ariane. This has riled Ariane's backers, who point to the huge government subsidies which finance shuttle flights. In 1983, a single shuttle mission of a week to ten days cost some $250 million.* This is just in operational costs and ignores the development bill for the spacecraft, a massive $15,000 million. By the time the shuttle's prices increase to bring them roughly into line with Ariane, the operating costs will probably have decreased,

* According to NASA estimates, this will reduce to some $125 million in 1984. NASA says that, by steadily reducing costs, each shuttle flight will on average break even by about 1988–9.

but it still appears that NASA will obtain, from customers whose satellites it is launching, only a fraction of its total overheads on the shuttle.

The absurdity of this state of affairs was illustrated in August 1983 when the eighth shuttle flight recouped for NASA a fee of a mere $8 million, or just over 3 per cent of the total bill for the trip into space. The fee was paid by the Indian government in return for the shuttle taking an Indian communications satellite into orbit. NASA says that the bill for operating the shuttle will decrease as the flight rate rises and engineers take advantage of the vehicle's inherent advantage over Ariane and other expendable launchers – its re-usability. According to NASA's plans, a fleet of four shuttles should, by the early 1990s, make forty trips into space a year, as compared to a total of four missions (by two shuttles) in 1983 and five in 1984.

However, this scheme could be wishful thinking. In April 1983, a panel of the US National Research Council told Congress that NASA may not have enough spares and other essential components (new engines, booster rockets and fuel tanks for example) to keep up the heavy flight programme that is envisaged. A maximum of eighteen launches a year may be all that is possible, said the panel. Even if by some miracle of financing (each shuttle costs about $1,000 million) NASA were able to increase its shuttle fleet from four to six, the annual flight rate would only with difficulty reach thirty-six, said the panel.

The uncertainty over running the shuttle throws into question

Table 19. Projection for satellites to be placed in orbit (1985–91)

Country/agency responsible	*Satellites*
Western Europe	43
US	135[1]
International agencies	24
Developed countries	25
Developing countries	31
Total	258

Source: Arianespace

1 Of these, 80 are military satellites or heavy vehicles that are not compatible with launch on Ariane.

how long the US government will be willing continually to subsidize its prestigious space vehicle. At some point, the shuttle's operator could well be forced to increase prices drastically, giving Ariane a clear advantage. At present, the balance between the two is fairly even. Arianespace says it aims to put into space about a quarter of the satellites that the non-communist world is due to launch between 1985 and 1991 (Table 19). It is on course to do this, even if the flight record (Table 20) for the rocket has been marred by two expensive and embarrassing failures. The shuttle's flights to the end of 1983 were all at least partial successes (Table 21), but then in February 1984 came a disastrous failure. The Challenger vehicle boosted two communications satellites, one owned by Western Union and the other by the Indonesian government, into incorrect orbits as a result of a malfunction in the upper stages, called a payload assist module (PAM) and made by McDonnell Douglas, that send satellites from the shuttle orbit to the geostationary ring. For months afterwards, engineers probed the reason for the failure and by midsummer they had still not gained a totally satisfactory explanation. The trouble seems to have been in the design of the engine nozzles in the two PAMs fitted to the pair of satellites.

Table 20. Ariane's launch record (to 31 December 1984)

Flight	*Date*	*Satellites orbited*
1	Dec. 1979	Test flight
2	May 1980	Failure (rocket crashed prior to putting in orbit 2 small satellites)
3	June 1981	Meteosat 2 (ESA); Apple (India)
4	Dec. 1981	Marecs (ESA)
5	Oct. 1982	Failure (rocket crashed before putting into orbit Marecs B for ESA)
6	June 1983	ECS-1 (ESA); Amsat radio satellite
7	Oct. 1983	Intelsat V (F7) for Intelsat
8	Mar. 1984	Intelsat V (F8) for Intelsat
9	May 1984	GTE-Spacenet-1
10	Aug. 1984	ECS-2 (ESA), Telecom-1 (France)
11	Nov. 1984	GTE-Spacenet-2, Marecs-132 (ESA)

Future flights: 7 planned for 1985; 9 in 1986. Arianespace expects it will launch 50–60 satellites in the five years 1986–90. Ariane can take a maximum of one satellite into orbit per flight.

Table 21. Space shuttle's launch record (to 31 May 1984)

Flight	*Date*	*Vehicle*	*Days in orbit*	*Satellites orbited (if any)*
1	Apr. 1981	Columbia	2	Test flight
2	Nov. 1981	Columbia	2	Test flight
3	Mar. 1982	Columbia	8	Test flight
4	June 1982	Columbia	7	DoD payload
5	Nov. 1982	Columbia	5	Anik C1 (Canada); SBS-3
6	Apr. 1983	Challenger	5	TDRSS 1[1]
7	June 1983	Challenger	6	Anik C2 (Canada); Palapa B1 (Indonesia)
8	Aug. 1983	Challenger	6	Insat 1B (India)
9	Nov. 1983	Columbia	10	Spacelab flight
10	Feb. 1984	Challenger	8	Westar;[1] Palapa B2[1]
11	Apr. 1984	Challenger	7	Repair of Solar Max
12	Aug. 1984	Discovery	6	SBS-4; Leasat; Telstar 302
13	Oct. 1984	Challenger	8	Science mission
14	Nov. 1984	Discovery		Anik; Leasat

1 Failed to enter correct orbit.

Future flights: 13 missions planned 1985, 17 in 1986, 24 in 1987, rising to 30 in 1990 and 40 in 1992 (1983 data). This assumes a total shuttle fleet of 4 (Challenger, Columbia, Discovery, and Atlantis).

Though NASA could fairly point out that this was McDonnell Douglas's responsibility, the incident hardly helped the credibility of the US shuttle fleet. NASA orbiters will be of little use to satellite-communications companies if the upper-stage motors used with the shuttles are unreliable.

Problems with the shuttle continued with the postponement by two months of the launch of Discovery, the third member of the shuttle fleet, after two aborted attempts to get it off the ground in June 1984. The incidents could only help the marketing efforts of Arianespace, which pointed out that expendable rockets are certainly less technologically sophisticated than the shuttle but are simpler and, on balance, less likely to go wrong. The point has not been lost on the Department of Defense in the USA, which is due to be a major customer for the shuttle but has started an investigation into the possibilities of placing contracts with expendable launchers such as Atlas Centaur (see below) as an insurance policy against further problems with the shuttle.

A major coup by Arianespace was in tempting two US satellite firms, GTE and SPCC, to put their spacecraft on the expendable launcher. By contrast, other US telecommunications companies, Western Union and SBS for instance, have stayed loyal to NASA and placed contracts with the shuttle. By mid-1984, Arianespace had firm orders for twenty-eight satellite launches, worth a total of £550 million.

Besides launching satellites owned either by ESA or European governments, Arianespace is due to take spacecraft for Brazil and Colombia into space. The shuttle is doing similar jobs for Indonesia, Canada and India. The countries in the Arab League have done their best to remain on good terms with everyone. They have booked both vehicles for the two Arabsat spacecraft that should provide TV and communications links in the Middle East.

Other contenders in the launch business are the Delta, Atlas Centaur and Titan rockets developed in the USA, which, like Ariane, are destroyed in the job of lifting their payloads into the heavens. These rockets, though operated now by the government, are to be sold to customers by private companies.* Eventually, the policy could extend to the shuttle. James Beggs, the administrator of NASA, says that his organization should become strictly a research and development agency, leaving operations of spacecraft to privately owned organizations. If the USA's vehicles were run by strictly commercial entities, of course, pricing policy could change dramatically. Table 22 sets out details for the world's launch vehicles.

Private industry in the USA also appears interested in developing totally new launch vehicles for satellites. One such company is Space Services of Houston, which plans flights with a rocket called Conestoga from 1986. The vehicle, based on rocket motors of the kind used in Delta, would fly from an island off the coast of

* In December 1983 and January 1984, NASA approved bids by Transpace Carriers to sell and operate the Delta. General Dynamics will do likewise with the Atlas Centaur. Both companies will have to lease launch facilities from the government. Martin Marietta is considering whether to take over operation of Titan from the Department of Defense.

Table 22. The world's launch vehicles

Vehicle	*Developer*	*Max. payload mass (tonnes)*		*Cost of putting satellite into geostat. orbit (1983 figures)*	
		Geostat. orbit	*Low Earth orbit*	*1.2 tonnes*	*2.4 tonnes*
Ariane[1]	CNES/ESA	2.4	—	\$24m[2]	\$50m[3]
Space shuttle[4]	NASA/Rockwell	7.5	30	\$16m[5]	\$20m[6]
Delta	NASA/McDonnell Douglas (US)	1.2	—	\$34m	—
Atlas Centaur	NASA/General Dynamics (US)	2.4	—	—	\$44m
Titan[7]	DoD/Martin Marietta (US)	4.5	13	—	\$100m
Proton	USSR	2–3	17	\$23m[8]	—
SLV-3[9]	India	—	0.1	—	—
N2[9]	Japan	0.8	—	—	—
Long March 3[9] (CSL X3)	China	0.9	—	—	—

Source: Comsat estimates, NASA, Arianespace, author's research.

1 Figures are for Ariane II/III. The more powerful Ariane IV, due in 1986, will lift 4.2 tonnes to the geostationary orbit.
2 Two satellites in one launch.
3 One satellite in a single launch.
4 Costs include extra sum for upper stage to boost satellites to the geostationary orbit.
5 Up to four satellites in one launch. Cost to increase to $26m in October 1985.
6 Up to two satellites in one launch. Cost to increase to $41m in October 1985.
7 Used mainly for military satellites.
8 Based on offer by Soviet Union in 1983 to launch satellites for Inmarsat.
9 Non-commercial launch vehicles.

Figures for costs are based on those at the end of 1983. They are approximate and depend on details of specific launches.

Mississippi. Alternatively, the real-estate developers who are the leading figures in Space Services are evaluating whether to launch from the island of Hawaii in the Pacific. Hawaii's position close to the Equator makes it an ideal site for rocket launches. Space vehicles ejected from the island would receive an extra 'kick' as a result of added angular velocity caused by the Earth spinning on its axis.

Another company making a similar proposal is the aptly named Starstruck, of Redwood City, California. It plans a series of launches of a small rocket called Dolphin, to be followed in the late 1980s by a more powerful vehicle called Constellation that would use similar technology. In what seems at first a hare-brained scheme, the company proposes to fire its rockets from out at sea, off the coast of San Diego. First, this makes it easier to comply with federal regulations that attempt to keep rocket launches away from built-up areas. Secondly, Starstruck thinks that with its 'floating rocket' scheme it could simply tow future launch vehicles to a position in the Pacific, on the Equator, to take advantage of the extra 'kick' when launching from this particular sector of the globe.

The Californian company's rockets would be hauled out to sea supported by cradles in a ship. They would be lowered in the water and launched vertically. The company insists that tests have shown this is perfectly safe and reliable. Starstruck's plans are unusual for another reason. Instead of using conventional technology, in which both oxidizer and fuel are liquids, the company's rockets would use a solid fuel and a liquid oxidizer. The fuel would be rubber compound, coated to the inside of the combustion chamber. The oxidizer would be liquid oxygen. The ignition of the oxygen would produce a reaction very similar to the burning of tyres, an everyday event on rubbish dumps all over the world. From this reaction would come a stream of hot gases, mainly carbon dioxide and water, to provide the rocket's thrust. Starstruck achieved its first successful test launch in 1984.

Still more imaginative is a scheme put forward by Bob Truax and Fell Peters, partners in another Californian venture called Truax Engineering. The company, based in Saratoga, had by early 1984 spent $600,000 on designing a rocket called Enterprise. Later, bigger versions of this would take satellites into orbit. Carrying on the rocket developers' tradition of choosing unusual names for their creations, Truax and Peters plan to call their future projectiles Excalibur and Sea Dragon.

To prove that it means business, Truax Engineering wants to put an astronaut into orbit on one of its rockets – after initial tests have shown that they are safe. Fell Peters himself says he will be that astronaut. The date of this launch is scheduled for around 1986. Peters is adamant that he will go down in history as the world's first 'private-enterprise' spacefarer.

On a more conventional level, the most significant event in 1983 in business operations in space was an unusual offer by the Soviet government. The USSR said it would sell to the West launches on its Proton rocket. The Proton is the heavy vehicle that takes Salyut space stations into orbit. The Soviet Union said it would launch satellites owned by Inmarsat, the international marine satellite organization in which the USSR has a 14 per cent stake. The suggested price is $23 million, undercutting Ariane, though rather more expensive than the space shuttle. Inmarsat was due to decide on the offer during 1984.

Some Western space experts expressed scepticism that the Soviet offer was genuine. They thought that customers might not be allowed near Soviet launch facilities – which are military installations out of bounds to foreigners. Thus a satellite's owner would not be permitted to watch his craft go into space – or even to check that launch preparations were adequate. Clearly, to show that the offer is realistic, the Soviets should straightaway open up their rocket sites to invited Westerners. The country could well have something to offer the satellite companies of Western Europe, the USA and Japan, and if the USSR really means what it says, it is the most startling reminder yet that the space business is truly becoming commercial.

10 ☆ The Future in Space

As mankind starts to flex its muscles in space, one illusion about exploiting the heavens should be dispelled. This is that, by exploring away from the Earth, people can get away from worldly problems. On the contrary, activities in space and on the Earth have never been more closely linked.

Many things of great importance to people in their everyday lives are tightly bound up with space ventures. That applies to communications, entertainment, business activities, national security – even weather forecasts. The major powers will gradually develop still bigger space projects, including factories to turn out new materials or even new energy systems. In future, the connections between the Earth and what goes on outside its atmosphere will grow stronger.

Much of the money spent on space technologies has benefited people on Earth. This is fairly obvious in projects for telecommunications, land mapping, meteorology and so on. Some of the beneficiaries are people in the poorer countries of the world, for whom remote sensing of natural phenomena or early warning of events such as hurricanes can mean the difference between life and death.

Less directly, spending on space technology can work its way through advantageously to other sectors of national economies. A 1979 study for the European Space Agency by the Louis Pasteur University in Strasbourg produced an impressive list of innovations that stemmed from work funded by the agency. These innovations were often in areas far removed from space technology itself. They included carbon fibres, computer cooling equipment, amplifier tubes, welding hardware, image receivers, techniques for measuring heat, thermal printers, microprocessors and new materials. According to the study, new ideas in areas such as these benefit the organizations doing space work, helping them to prosper and to generate wealth and employment.

A similar picture is painted in an American document called *Spinoff 79*. This is a report by NASA of the way that space research produces innovations in other areas. As examples that have arisen from the space programme in the USA, the report lists advances in waterflow measurement, ship navigation, oil pipelines, window insulation, infrared heaters, solar energy, dentistry, checking of blood pressure, smoke detection, sports equipment and cable cars.

Some studies have attempted to quantify the benefits to a national economy through such research spending. In 1971, the Midwest Research Institute of the USA studied NASA's activities. It said that spending by the space agency of $29 billion in the ten years to 1969 would generate a total return to the economy of $207 billion by the year 1987. That translates into a 7 to 1 return on investment or an estimated 33 per cent discounted rate of return.

In 1975, another organization, Mathematica, looked at four techniques in whose development NASA has played a role. The researchers concluded that, by 1984, $7 billion worth of benefits to the US economy could be attributed to NASA's development of cryogenic insulation, gas turbine engines, integrated circuits and computer modelling.

NASA tries to push its ideas into the economy by a system of patent waivers. By this, companies can take over the rights to NASA inventions with a minimum of bureaucracy. Of 860 inventions passed into industry from 1961 to 1975, some 20 per cent resulted in successful products or industrial processes, according to a NASA study. The most successful rates of commercialization were in transportation, electrical machinery and metals. Of a total of 3,512 patents granted to NASA between 1959 and 1979, 523 led to licence deals with individuals or companies in which a product or process was commercialized.

Forecasters are generally confident about the development of space commercialization. In May 1983, Paul Keaton of the Los Alamos National Laboratory published a paper entitled 'Why Billions Could and Should be Spent on Space' in the American journal *Astronautics and Aeronautics*. Keaton argued that the USA should spend $130 billion on space projects over the twenty-two years to 2005. Such a programme would boost the economy as well as accomplishing specific goals – developing new manufacturing processes for example. 'The dangers of not making major long-term commitments are twofold,' wrote Keaton. 'An unnecessarily miserly approach to our nation's space programme can be a self-fulfilling prophecy, and without a long-range perspective we run the risk of spending the money unwisely in fits and starts.'

Dr Klaus Heiss, a champion of private-enterprise ventures in space, has produced a blueprint for the different stages in space exploitation. Dr Heiss is best known for trying to buy a private space shuttle for $1 billion, a project that petered out in 1983. Speaking at a conference in Strasbourg in 1980, Heiss pinpointed five phases in the likely evolution of space industrialization:

(1) *Information phase* (1960s to 1990) People use space to gather, transmit and evaluate information on a worldwide basis. This stage features projects in communications, remote sensing, ocean scanning, military surveillance, climate monitoring. Researchers also study the scientific aspects of space.

(2) *Large structures* (1980s to 2010) With the space shuttle, the USA will construct, in low orbits, large assemblies for manufacturing and energy production. Engineers will also build large 'antenna farms' in space for relaying signals between different parts of the Earth. People will experiment with robots and other automated hardware. This is a learning phase in which engineers and scientists understand more deeply the factors involved in living and working in a state of weightlessness.

(3) *Industrial prototypes* (1990s to 2025) The first working models of space factories will appear. Much of this activity will be organized by governments, with substantial industrial collaboration.

(4) *Industrial phase* (2010s to 2075) Private space projects will sustain themselves without the support of governments. For instance, electricity utilities will put their own solar power systems into space. Other activities will include waste disposal, enzyme production, crystal growth, materials processing, closed agricultural systems.

(5) *Space habitation* (2075 onwards) The threshold stage in space colonization. People will use space not just for important industrial processes but as a place to live. 'After the crossing of this threshold,' says Heiss, 'no physical resource limitations seem likely to limit the growth of such communities from then on for thousands of years.'

Other writers have tried to predict how people will move into space on a large scale. Stephen Dole did a study on this for the RAND Corporation in the USA in the 1960s. In a book based on this study, written with Isaac Asimov and called *Planets for Man*, Dole explained:

> As it is now, one planet, the Earth, supports the entire human race. One planetary catastrophe could completely destroy us. But if the human race were living on a number of planets scattered around the galaxy, its immortality would be assured. What's more, the opportunities for variations in culture and in outlooks would be vastly multiplied, and the interactions among the multitude of human subgroupings might vastly accelerate the overall progress of mankind toward ultimate mastery of the physical universe.

The book's self-confident tone may jar with many people wrestling with the purely terrestrial problems of the 1980s. To attain immortality for mankind does not seem particularly worthy. Far better to make a reasonable life on Earth, in the short term, for as many of its inhabitants as possible. 'Ultimate mastery' of the universe sounds a little strident, too. *Planets for Man* continues in a similar vein.

> Always, those volunteering for the next expedition into the unknown (in space habitation projects) will tend to be adventurous, self-reliant, inquisitive, courageous and hardy pioneers, while those selected to go will be chosen on the basis of good health, high professional competence, emotional stability, reliability of judgement and so on. In the main, these characteristics will be passed on to their descendants so that a kind of selection process will take place, with those at the frontier of the wave through the galaxy always representing some of the best qualities of mankind, and leavening all of mankind with these qualities.

The aggressive approach to space engineering that is seen in Dole's book was in step with the attitudes of the 1960s. It also fits in with some of the ideas that people today have about space projects. In particular, it equates with the views of some business corporations that are going all-out to harness space for commercial ends, and with those of military chiefs in the USA and USSR who look upon the heavens as a place where battles may be fought in future wars.

Daniel Deudney, a researcher at the Worldwatch Institute in

Washington DC, attacks what he calls the 'lawlessness and the escapism of the frontier mentality' in space affairs. In a booklet in 1982, he wrote:

> The superpowers have a first-come, first-served attitude toward a global medium that is no more their own than the air they breathe or the ocean straits through which they demand right of passage. Those with a positive space programme succumb to the old frontier illusion that an Eden of abundance and harmony awaits us in space. The urge to pick up and move to a new land when things start getting bad in the old country has taken on a new high-technology character.

Deudney goes on:

> The advocates of war in space are similarly escapist. Recognising that war on Earth is too horrible to contemplate, they want to ship it out into space where it will not hurt anyone, a position that has a surprisingly diverse group of supporters. This image of space war being like an eighteenth-century battle, when armies fought other armies while civilian life went on largely unaffected, does not bear up to examination. The military space activities of the superpowers are too interconnected with the command, targeting and warning systems of strategic nuclear weapons on Earth. War in one realm would almost surely become a war in the other.

One of the biggest tasks in space projects is to convert technology developed for military purposes to peaceable tasks. One good example is navigation satellites. These were devised so that soldiers or military vehicles could pinpoint their positions in battles. The satellites can even direct guided missiles to their targets. A big challenge is to make the information from these space vehicles available to civilians – for example in airline navigation or even to help hikers to find their way in unfamiliar countryside. Such a transfer may well happen with the USA's new generation of military spacecraft for navigation, the Global Positioning System.

Another, still more visionary goal is to set up an international system of space surveillance using the 'spycraft' technologies developed by the superpowers. Satellites would scan the Earth for weapons – missile sites and so on. The craft would spot the installation of hardware prohibited by arms-control accords and would act as 'flying policemen', passing on information about transgressors to the relevant international bodies, for instance at the United Nations.

In 1978, France mooted the idea of an international satellite monitoring agency. Since then, the proposal has largely been forgotten, though if resurrected it could be a useful way of reducing the threat of nuclear conflict.

Several topics will be of great importance over the next few years in space activities. They include the overcrowding of the geostationary orbit and discussion on how to share out this precious resource in a fair way. Space planners will debate whether expendable rockets or re-usable shuttles form the most economic launch vehicles. They will discuss whether governments or private companies should play the biggest role in space ventures, and the discussion as to whether satellite systems or new cable networks offer the most efficient communications links will continue.

Superimposed on these issues is one of overriding importance. That is how to strike the right balance between big, adventurous projects that capture the imagination and the smaller schemes that may not seem so exciting but which may do more to meet people's real needs.

Do we, for example, really require large, expensive, permanently manned space stations that do jobs of doubtful benefit to anyone? Or should we concentrate on simple satellite networks that relay to Earth the type of information – in remote sensing for example – that we know will be useful?

Do training schemes for Third World space scientists represent better value for money than projects on nuclear propulsion? Could engineers really produce such large and complex edifices as solar power satellites without something going horribly wrong? Instead of spending vast sums on advanced communications satellites, would it be better to put the money towards telecommunications equipment for the hundreds of millions of people who have never seen a telephone?

The answers to these questions will become clearer over the next quarter of a century. For the moment, we should heed the conclusion drawn by Daniel Deudney:

> Space's most important lesson is its reaffirmation that humanity's fate will be decided on Earth. Colonisation of space in the distant future is something scientists can speculate and science fiction writers can imagine. But for all our looking and probing of the Universe, we have yet to find any place as habitable as the remotest, most forbidding parts of this planet.

Space exploration has taught us just how rare and precious the Earth is. Knowing how little there is in space gives us all the more reason to preserve the fullness of life on Earth.

Further Reading

Chapter 1: The Exploitation of Space

Irving Adler, *Seeing the World from Space*, Signet, 1962.
Isaac Asimov, *The Universe*, Penguin, 1977.
Wernher von Braun, *First Men to the Moon*, Frederick Muller, 1961.
Ralph Chipman (ed.), *The World of Space* (prepared for United Nations space conference, UNISPACE 1982), Prentice-Hall, 1982.
Arthur C. Clarke, *The Exploration of Space*, Harper, 1951.
Arthur C. Clarke, *Profiles of the Future*, Pan, 1964.
A. Dattner, *Reflections on Europe in Space*, ESA, 1982.
Giovanni Dondi, *Space Activities in the Eighties*, ESA, 1981.
Frederick Durant (ed.), *Between Sputnik and the Shuttle*, American Astronautical Society, 1981.
Kenneth Gatland (principal author), *The Illustrated Encyclopedia of Space Technology*, Salamander, 1981.
Richard Lewis (principal author), *Space Exploration*, Salamander, 1983.
Patrick Moore, *Space in the Sixties*, Penguin, 1963.
L. Napolitano (ed.), *Space: Mankind's Fourth Environment*, Pergamon, 1982.
Hermann Oberth, *Man into Space*, Weidenfeld and Nicolson, 1957.
Office of Technology Assessment, *Civilian Space Policy and Applications*, US Congress, US Government Printing Office, 1982.
Office of Technology Assessment, *Civilian Space Stations and the US Future in Space*, US Government Printing Office, 1984.
Geoffrey Pardoe, *The Future for Space Technology*, Frances Pinter, 1984.

Stephen Pizzey, *Exploring the Solar System*, HMSO, 1977.
H. G. Wells, *A Short History of the World*, Penguin, 1977.
Europe: Two Decades in Space, recollections by some of the principal pioneers, ESA, 1984.
Space 25, special issue of *Spectrum*, Institute of Electrical and Electronics Engineers (US), 1983.
Twenty Years of European Cooperation in Space, ESA, 1984.

Chapter 2: The Propulsion Pioneers

Carlisle Adams and Wernher von Braun, *Careers in Astronautics and Rocketry*, McGraw-Hill, 1962.
Wernher von Braun and Frederick Ordway III, *History of Rocketry and Space Travel*, Nelson, 1967.
Walter Dornberger, *V-2*, Hurst and Blackett, 1954.
Kenneth Gatland, *Missiles and Rockets*, Blandford, 1975.
V. P. Glushko, *Development of Rocketry and Space Technology in the USSR*, Novosti Press, 1973.
Robert Goddard, *Rocket Development*, Spectrum, 1961. (Condensation of Goddard's notebooks.)
Ernst Klee and Otto Merk, *The Birth of the Missile: The Secrets of Peenemünde*, Harrap, 1965.
Milton Lehman, *This High Man (The Life of Robert Goddard)*, Farrar and Straus, 1963.
Willy Ley, *Rockets and Space Travel*, Viking, 1947.
Frederick Ordway and Mitchell Sharpe, *The Rocket Team*, Heinemann, 1979.
Albert Parry, *Russia's Rockets and Missiles*, Macmillan, 1960.
Nels Parson, *Missiles and the Revolution in Warfare*, Harvard University, 1962.
G. Edward Pendrey, *The Coming Age of Rocket Power*, Harper, 1947.
John Taylor, *Rockets and Satellites Work Like This*, Phoenix, 1968.
Konstantin Tsiolkovsky, *Beyond the Planet Earth*, Pergamon, 1960 (originally published 1920).
Spaceflight Technology, *Proceedings of First Commonwealth Spaceflight Symposium*, Academic Press, 1960.

Chapter 3: Communications from the Sky

Eric Burgess, 'Into Space', *Aeronautics*, November 1946.
Eric Burgess, *Satellites and Spaceflight*, Chapman and Hall, 1957.
Arthur C. Clarke, 'Extra Terrestrial Relays', *Wireless World*, October 1945.
Max Faget, *Manned Space Flight*, Holt, Rinehart and Winston, 1965.
A. T. Lawson, *Window in the Sky*, David and Charles, 1979.
M. Vassilev and V. V. Dobronravov, *Sputnik into Space*, Souvenir Press, 1958.
An Introduction to Space, HMSO, 1966.
Soviet Writings on Earth Satellites and Space Travel, various authors, MacGibbon and Kee, 1959.

Chapter 4: Battle of the TV Beams

J. Chaplin, *Developments in LW Power Satellite Transponder Leasing for TV Distribution*, Council of Europe, 1984.
Mark Long and Jeffrey Keating, *The World of Satellite Television*, Book Publishing Company, 1983.
C. Rosetti and J. Chaplin, *A European View of Direct Broadcasting by Satellite*, ESA, 1982.
Direct Broadcasting by Satellite, Symposium Report, Royal Aeronautical Society, 1982.
Direct to Home Satellite Broadcasting for Canada, Canadian Department of Communications, 1982.

Chapter 5: The Satellite Snoopers

W. Collins and J. Van Genderen (eds), *Remote Sensing Applications in Developing Countries*, Remote Sensing Society, 1978.
John McElroy, *Earthview: Remote Sensing of the Earth from Space*, NOAA, 1984.
Stephen Pizzey, *Exploring Remote Sensing*, HMSO, 1977.
Commercialisation of Civil Remote Sensing, Report by Comsat, 1982.
Global Change: A Biogeochemical Perspective, Report of Workshop Chaired by Michael McElroy, NASA, 1983.

Global Change: Impacts on Habitability, Report of Workshop Directed by Richard Goody, NASA, 1982.
Remote Sensing Applications for Environmental Studies, Report of Symposium, ESA, 1983.
Remote Sensing and Digital Mapping, Report by House of Lords Select Committee on Science and Technology, HMSO, 1983.
Resource Sensing from Space: Prospects for Developing Countries, US National Academy of Sciences, 1977.
Satellites in Meteorology, Oceanography and Hydrology, World Meteorological Organization, 1982.
Satellite Remote Sensing for Developing Countries, Report of Symposium, ESA, 1982.
Toward an International Geosphere-Biosphere Program, US National Academy of Sciences, 1983.

Chapter 6: Space War

M. Callaham and K. Tsipis, *High Energy Laser Weapons – A Technical Assessment*, Massachusetts Institute of Technology, 1980.
Kenneth Gatland, *Development of the Guided Missile*, Iliffe, 1954.
Bhupendra Jasani (ed.), *Outer Space – A New Dimension of the Arms Race*, Taylor and Francis, 1982.
Nels Parson, *Missiles and the Revolution in Warfare*, Harvard University, 1962.
Curtis Peebles, *Battle for Space*, Blandford, 1983.
Stockholm International Peace Research Institute, *World Armaments and Disarmaments, SIPRI Year Book 1983*, Taylor and Francis, 1983.
The Technology of Military Space Systems, International Secretariat of the North Atlantic Assembly, 1982.

Chapter 7: Space Work

Joseph Bell, *Seven into Space*, Ebury Press, 1960.
Hilding Bjurstedt (ed.), *Biology and Medicine in Space*, ESA, 1979.
Arthur C. Clarke and the editors of *Life*, *Man in Space*, Time-Life, 1968.
Henry Cooper, *Apollo on the Moon*, Dial Press, 1969.

Bevan French, *The Moon Book*, Penguin, 1977.
Tim Furniss, *Manned Space Log*, Jane's, 1983.
Yuri Gagarin, *Road to the Stars*, Foreign Languages Publishing House (Moscow), 1962.
Kenneth Gantz (ed.), *Man in Space*, Hollis and Carter, 1960.
W. F. Hilton, *Manned Satellites*, Hutchinson, 1965.
Beirne Lay, *Earthbound Astronauts*, Prentice-Hall, 1971.
Robert Naumann and Harvey Herring, *Materials Processing in Space: Early Experiments*, NASA, 1980.
James Oberg, *Red Star in Orbit*, Harrap, 1981.
Philip Pierce and Karl Schuon, *John H. Glenn – Astronaut*, Harrap, 1962.
Francis Riley and Douglas Sailor, *Space Systems Engineering*, McGraw-Hill, 1962.
Peter Ryan, *The Invasion of the Moon 1969*, Penguin, 1969.
Peter Smolders, *Soviets in Space*, Lutterworth, 1973.
The Seven Mercury Astronauts, *The Astronauts: Pioneers in Space*, Cassell, 1961.
Extraterrestrial Materials Processing and Construction, Report by Lunar and Planetary Institute (US), 1980.
Materials Sciences under Microgravity, Proceedings of ESA Symposium, ESA, 1983.
Results of First US Manned Orbital Space Flight, NASA, 1962.
Spacelab Data Book, ESA, 1983.
The Astronauts Book, various authors, Panther, 1966.

Chapter 8: Enter the Robot

D. Akin, M. Minsky, E. Thiel and C. Kurtzman, *Space Applications of Automation, Robotics and Machine Intelligence Systems*, NASA, 1983.
Oleg Gazenko, *Man in Space* (lecture), Institute of Biomedical Problems (Moscow), 1982.
Peter Marsh, *The Robot Age*, Abacus, 1982.
John von Neumann, *Theory of Self-Reproducing Automata*, University of Illinois, 1966.
Mitchell Sharpe, *Living in Space*, Aldus, 1969.
Georg von Tiesenhausen and Wesley Darbro, *Self Replicating Systems: A Systems Engineering Approach*, NASA, 1980.

Charles Wunder, *Life into Space: An Introduction to Space Biology*, F. A. Davis, 1966.
Advanced Automation for Space Missions, A Technical Summary, University of Santa Clara, 1980.
Machine Intelligence and Robotics, Report of NASA Study Group, NASA, 1980.

Chapter 9: Rockets: Technology and Competition

Howard Allaway, *The Space Shuttle at Work*, NASA, 1979.
Donald Cox and Michael Stoiko, *Spacepower: What it Means to You*, John Winston, 1958.
D. G. Fearn, *Solar Power Satellites; A Review of the Space Transportation Options*, Royal Aircraft Establishment, 1980.
D. G. Fearn, *A Review of Future Orbit Transfer Technology*, Royal Aircraft Establishment, 1981.
Kenneth Gatland, *Astronautics in the Sixties*, Iliffe, 1962.
Peter Glaser, *Evolution of the Solar Power Satellite Concept*, Symposium of Energy from Space, UN Space Conference, Vienna, 1982.
Adrian Mann, *Spadeadam*, Parsons Memorial Lecture, 1963.
Space Shuttle Transportation System, Rockwell International, 1981.
The Solar Power Satellite Concept, NASA, 1979.
13th Report of the Estimates Committee on Space Research and Development, House of Commons, 1967.
5th Report of the Select Committee on Science and Technology, House of Commons, 1971.

Chapter 10: The Future in Space

George Brown and Timothy Lynch, *A Review of the Proposal for an International Satellite Agency*, 1984 (private paper).
Daniel Deudney, *Space: The High Frontier in Perspective*, Worldwatch, 1982.
Stephen Dole and Isaac Asimov, *Planets for Man*, Methuen, 1965.
H. Lorber and R. Drakem, *The Economic Benefits of Space Development*, Los Alamos National Laboratory, 1982.

Economic Effects of Space and Other Advanced Technologies, Report of Workshop, ESA, 1980.
Spinoff 79, NASA, 1979.
Technical Innovations and Transfers of Technology: The Role of ESA, Study under direction of Prof. Jean Paul Fitoussi, Université Louis Pasteur, 1979.

Index

MORE ABOUT PENGUINS, PELICANS AND PUFFINS

For further information about books available from Penguins please write to Dept EP, Penguin Books Ltd, Harmondsworth, Middlesex UB7 ODA.

In the U.S.A.: For a complete list of books available from Penguins in the United States write to Dept DG, Penguin Books, 299 Murray Hill Parkway, East Rutherford, New Jersey 07073.

In Canada: For a complete list of books available from Penguins in Canada write to Penguin Books Canada Ltd, 2801 John Street, Markham, Ontario L3R 1B4.

In Australia: For a complete list of books available from Penguins in Australia write to the Marketing Department, Penguin Books Australia Ltd, P.O. Box 257, Ringwood, Victoria 3134.

In New Zealand: For a complete list of books available from Penguins in New Zealand write to the Marketing Department, Penguin Books (N.Z.) Ltd, Private Bag, Takapuna, Auckland 9.

In India: For a complete list of books available from Penguins in India write to Penguin Overseas Ltd, 706 Eros Apartments, 56 Nehru Place, New Delhi 110019.

THE PENGUIN ENGLISH DICTIONARY

The Penguin English Dictionary has been created specially for today's needs. It features:

* More entries than any other popularly priced dictionary
* Exceptionally clear and precise definitions
* For the first time in an equivalent dictionary, the internationally recognised IPA pronunciation system
* Emphasis on contemporary usage
* Extended coverage of both the spoken and the written word
* Scientific tables
* Technical words
* Informal and colloquial expressions
* Vocabulary most widely used *wherever* English is spoken
* Most commonly used abbreviations

It is twenty years since the publication of the last English dictionary by Penguin and the compilation of this entirely new *Penguin English Dictionary* is the result of a special collaboration between Longman, one of the world's leading dictionary publishers, and Penguin Books. The material is based entirely on the database of the acclaimed *Longman Dictionary of the English Language.*

1008 pages   051.139 3   £2.50 ☐

A CHOICE OF PELICANS AND PEREGRINES

A CHOICE OF PELICANS AND PEREGRINES

☐ ***A Question of Economics*** **Peter Donaldson** £4.95

Twenty key issues – from the City and big business to trades unions – clarified and discussed by Peter Donaldson, author of *10 × Economics* and one of our greatest popularizers of economics.

☐ ***Inside the Inner City*** **Paul Harrison** £4.50

A report on urban poverty and conflict by the author of *Inside the Third World*. 'A major piece of evidence' – *Sunday Times*. 'A classic: it tells us what it is really like to be poor, and why' – *Time Out*

☐ ***What Philosophy Is*** **Anthony O'Hear** £3.95

What are human beings? How should people act? How do our thoughts and words relate to reality? Contemporary attitudes to these age-old questions are discussed in this new study, an eloquent and brilliant introduction to philosophy today.

☐ ***The Arabs*** **Peter Mansfield** £4.95

New Edition. 'Should be studied by anyone who wants to know about the Arab world and how the Arabs have become what they are today' – *Sunday Times*

☐ ***Religion and the Rise of Capitalism***
R. H. Tawney £3.95

The classic study of religious thought of social and economic issues from the later middle ages to the early eighteenth century.

☐ ***The Mathematical Experience***
Philip J. Davis and Reuben Hersh £6.95

Not since *Gödel, Escher, Bach* has such an entertaining book been written on the relationship of mathematics to the arts and sciences. 'It deserves to be read by everyone . . . an instant classic' – *New Scientist*